D1104259

WHY HELL STINKS OF SULFUR

Why Hell Stinks of Sulfur

Mythology and Geology of the Underworld

Salomon Kroonenberg

Translated by Andy Brown

REAKTION BOOKS

For my grandchildren

Published by
Reaktion Books Ltd
33 Great Sutton Street
London EC1V 0DX, UK

www.reaktionbooks.co.uk

This publication has been made possible with financial support
from the Dutch Foundation for Literature

Printed and bound in China
by C&C Offset Printing Co., Ltd

British Library Cataloguing in Publication Data

Kroonenberg, Salomon Bernard.
Why hell stinks of sulfur: mythology and geology of the underworld.
1. Earth – Internal structure – Popular works.
2. Earth – Mythology.
3. Hell in literature.
I. Title
551.1'1-dc23

ISBN 978 1 78023 045 0

CONTENTS

❦ ONE ❦

The Gobstopper

So dark and deep and nebulous it was,
try as I might to force my sight below
I could not see the shape of anything . . .
Dante Alighieri, *Inferno*, IV:10–12

Why did astronomers get to study heaven, and we geologists hell? They can look into space for billions of light years, while we can't even see the moles that destroy the lawns beneath our feet. They go to Mars, but we don't go to the centre of the Earth. Governments spend millions on manned space flight but astronauts have to pay for their own burials. Lovers look up at the Moon, while convicts stare down at the ground. In Paradise we find eternal happiness and in hell the most miserable of deaths. Why do we geologists always come off worst?

I know why: because heaven is transparent and the Earth is not. Evolution has not given us eyes that allow us to see through rock. And in the darkness, people get scared. That's why hell is underground. If the Earth were transparent, like a giant glass marble, they would spend all day lying flat on their stomachs and looking down at all the activity below them. They would see how moles row laboriously through the soil with their little pink hands. They would see the seeds fighting to be the first to stick their heads above ground in the spring: lungwort, ground ivy and wood anemones. And they would hold a competition to see who could peer the furthest into the depths. People would find the Earth so beautiful that they would want to be cremated rather than buried, so that the soil would remain clean. And they would want to travel to the Earth's core.

But the Earth is not transparent. Most of the world's cultures believe that beneath the surface lies the Underworld, the kingdom of the dead. And yet that underground world contains so much that is beautiful: sparkling ores and metals, magnificent yellow sulfur crusts,

A one-inch gobstopper, whole and in cross-section.

blue sapphires, red cinnabar, green malachite, razor-sharp, metres-long gypsum crystals, dripstone caves, subterranean rivers, fragile shells from the dawn of evolution, and the giant bones of extinct monsters.

For geologists, the interior of the Earth is like the inside of a gobstopper, one of those old-fashioned, hard sweets that you can hardly fit into your mouth. On the outside they are white with coloured speckles, but if you suck on one for a while, when you take it out of your mouth, it will be a different colour: blue, green or pink. Each time you suck on it (don't bite it!), you discover a new layer. I bought a few gobstoppers and asked the technician in our lab to cut one through the middle, just as he does with rocks when preparing them for the microscope. It's a perfect representation of the concentric structure of our globe.

It took a long time to discover that the Earth is just like a giant gobstopper. It was not that people had never given any thought to what it looked like under the ground, but myth and science, legend and observation, fear and curiosity are all mixed up together in a fascinating way. None of the heroes I learned about at secondary school who had been to the Underworld and returned ever recounted anything about what it was like. Odysseus, Theseus, Orpheus, Heracles, Aeneas, Dante and many others went down there, usually because of a woman or sometimes just to seek advice or out of curiosity, but they came back without saying anything about all the wonderful things they had seen or what they had learned. They told of seeing sinners splashing around in rivers of pitch, but did not say where the pitch came from. They saw Lucifer stuck in the ice, but did not explain why it was so

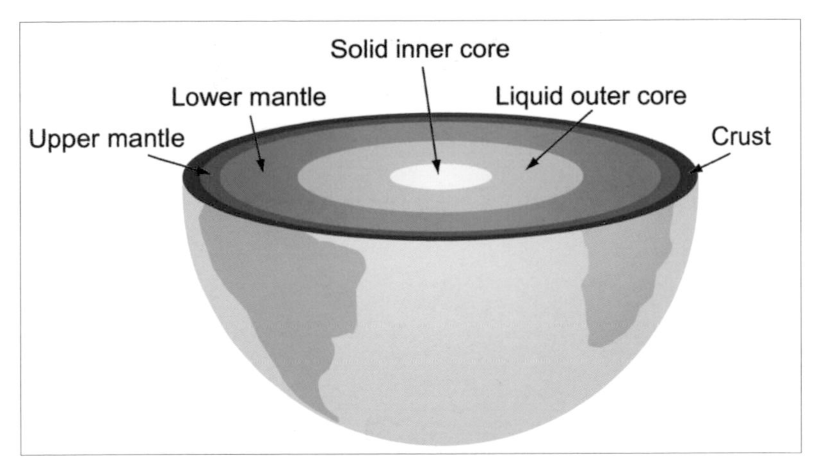

The Earth as a gobstopper as modern geologists see it.

cold. Gustave Doré (1832–1883) drew the most fantastic landscapes of Dante's Inferno, but how were those landscapes created? Was it limestone down there, or basalt? And why does hell stink of sulfur?

It is time someone took a look at hell from a geological perspective. So I'm going down there to do fieldwork, armed with a hammer and compass, in the footsteps of my heroes. Once I've found the entrance, I'll descend into hell, a little deeper in each chapter, roughly following the circles of Dante's hell. And I'll give you a detailed report of my journey – as long as I succeed in getting out again – not because of a macabre sort of *oltretomba* tourism, but to show how fear and imagination have triumphed over science, and how much imagination we need even now to picture what it must look like down there.

After all, the subsurface remains the most unknown part of our planet, despite the fact that the centre of the Earth is no further away than London is from Chicago. It is also high time to take a closer look at the world beneath our feet, which we drill so many holes in, it's a wonder there's any ground left at all. We suck the Earth dry and hollow it out, and fill the gaps with everything and anything we want to hide from the light of day. We store radioactive waste in the ground, while the Earth itself has produced natural nuclear reactors. We pump our last remaining carbon dioxide into empty gas fields, while life itself has deposited 90 per cent of all the original CO_2 in the Earth's crust in the form of limestone.

And why do we do that? Because the air has to be clean, we say. The water has to be clean and the soil has to be clean. But doesn't the

Dante's Inferno as portrayed by Stradanus (Jan van der Straet, 1587): was it limestone?

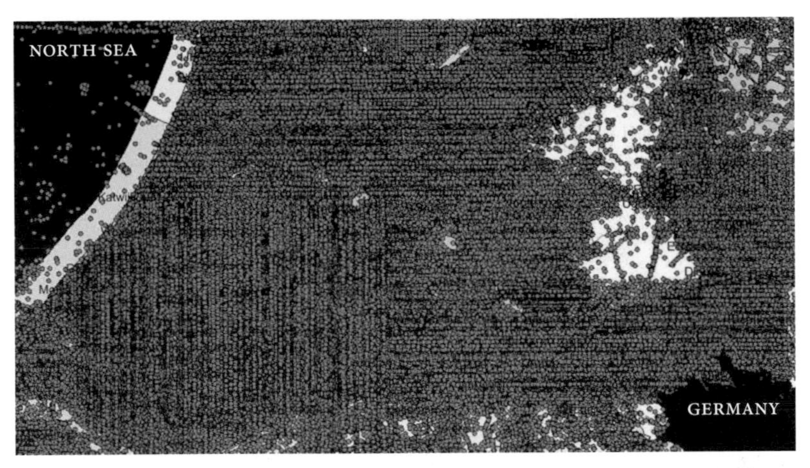

Density of boreholes in the central Netherlands, according to the DINO database.

subsurface have to be clean? Of Empedocles' four elements of earth, water, fire and air, it is earth that always draws the short straw.

The history of the Earth is written in the subsurface. It tells of unimaginable extremes in our climate; of devastating earthquakes, eruptions and floods; of the evolution of the human race. Everything we know about our ancestors – about Neanderthals, about *Homo erectus* – we know from excavated skeletons. You can't find the history of the atmosphere in the atmosphere itself: carbon dioxide molecules only remain in the air for a couple of thousand years before being reabsorbed by the oceans or by plants. The history of the atmosphere is recorded in the subsurface: in the sediments of the deep seas, in deep peat bogs and lakes, in the air bubbles in the ice caps of Greenland and Antarctica. You can't understand heaven without knowing hell. The Earth is its own history book.

Perhaps we should no longer see the world beneath our feet as a black box in which to dig tunnels, a supermarket for raw materials, a carpet to sweep our rubbish under, or the final resting place of the dead, but as an irreplaceable archive and a living ecosystem whose riches we have as yet hardly been able to fathom.

❧ TWO ❧

Jerusalem

The sun was touching the horizon now,
the highest point of whose medium arc
was just above Jerusalem;
Dante Alighieri, *Purgatorio*, II:1-3

There is no better place to start my geological descent than Jerusalem, the place from which Christ himself descended into hell. Christ's journey to the Underworld is not described in the Bible, but in the Gospel of Nicodemus, an apocryphal work that was rejected for the New Testament by the strict hand of an editor. I would have left it in, as it gives a lively and dramatic portrayal of hell that is to be found nowhere else in the Bible.

There was a rumour that the two sons of the high priest Simeon, Karinus and Leucius, had risen from the dead. The high priests who had accused Christ, including Annas, Caiaphas, Nicodemus, Joseph and Gamaliel, sought out the place where they had been buried and found that their tombs were indeed empty. The two dead men proved to be walking around again in the city of Arimathea. The high priests found them, took them to the synagogue, and locked the door behind then. Then they asked how it was possible that they had returned from the dead. But Karinus and Leucius made the sign of the cross on their lips with their fingers to indicate that they were not allowed to say anything. The high priests asked whether it was Jesus who had raised them from the dead. Then Karinus and Leucius asked for a sheet of paper and wrote their story down.

'After Jesus died on the cross and was laid in his tomb,' they wrote,

> he restored many of the dead to life. We were sitting with our forefathers in the darkness when suddenly the golden heat of the sun, a purple and royal light, started to shine. Esaias said that it was the light of the Father, the Son of God. And Seth, the

Christ's descent into hell.

son of the patriarch Adam, told us that he had heard from the archangel Michael that when five thousand and five hundred years had passed, the Son of God would come to the Earth and bring the bodies of Adam and of all the dead to Paradise.

We also saw that Satan and Hades argued about the coming of Jesus. Satan complained that Jesus had healed people that he, Satan, had made blind, lame, dumb or leprous. He had stirred up the Jewish people against Jesus, thrust a spear in his side, gave him gall and vinegar to drink, and nailed him to a cross. But Hades said that Satan was much too powerful to be troubled by Jesus and that he had lost dead souls before, including Lazarus. 'But that was Jesus, too!' Satan protested.'

While they were talking, we suddenly heard a thunderous voice: 'Remove your everlasting gates, the King of Glory shall come in.' When Hades heard that, he said to Satan: 'Go and fight against him, if you're so powerful!' And he ordered his servants to shut the brass gates and secure them with bars of iron, to stop Jesus from getting in. But David, Esaias and all the saints cried: 'Open the gates!' And the Son of God appeared in the form of a man and lightened the eternal darkness and broke the unbreakable bonds. Hades and his servants were afraid and asked: 'Who are you? Are you perhaps Jesus, who

> Satan said would be the ruler of the whole world?' Then the King of Glory trampled on death, took hold of Satan and delivered him to Hades, saying: 'From now on Satan shall be in your power, in exchange for Adam and his children.' Then Jesus led Adam and all the saints out of hell. They were allowed to spend three days on Earth to testify that Christ had indeed risen from the dead, and then the archangel Michael took them to Paradise. But we, Karinus and Leucius, had to stay in Arimathea and pray.

Then Karinus and Leucius handed their written accounts to the high priests, at which point they turned as white as chalk, faded away and disappeared. But their writings remained.

How did Jesus enter hell? There is no account of that anywhere. After he was crucified, his body was placed in a cave in Golgotha that Joseph of Arimathea had ordered to be hacked out of the rock for himself. Joseph had rolled a large stone in front of the entrance to the cave to ensure that the Jews did not give the body to the vultures. But there was an earthquake, the stone rolled away, and the body disappeared. That must have been the moment that Jesus was in hell. But where was the entrance? In the back of the cave? That is why I am here. The Old City is divided into four quarters: Jewish, Christian, Muslim and Armenian. Each religion has its own hell. I walk into the Christian district through the Jaffa Gate, which is little more than a small opening in the metres-thick white, yellow and grey limestone wall that encircles the Old City. I pick my way through narrow alleys and gateways and up and down steps. Electricity lines, washing lines and cables with lamps hang between the houses like limp spiders' webs. And then I see it: the Church of the Holy Sepulchre, where the tomb of Jesus is said to lie. I go inside.

But unfortunately the church has been built rather unceremoniously on top of the tomb and there is no longer anything to suggest that it was once a cave: the overhanging roof of the cave was removed when the church was first constructed in the fourth century. Since then the Persians, the Arabs and fire have all done their bit to ensure that little remains that is original. Now there is a nondescript sepulchre, the size of a shipping container, that could have been anybody's and which, through lack of maintenance, is fixed to the floor with cables. Nor is there a stone that can roll away at the next earthquake. Even the hill of Golgotha itself was radically hacked away for the building of the church. All that can be seen is a glimpse of cracked limestone through a small, inconspicuous window. But there are no signs of a cave. If I were a believer in any of the faiths that should have been responsible for maintaining the tomb, I would have cried 'Sacrilege!' But now I cry 'Destruction of a unique natural monument!' If Christ entered hell though this cave, all traces have been thoroughly erased. Incense has triumphed over sulfur.

I continue my walk through Jerusalem. At the Damascus Gate in the Muslim district on the north side of the city, I climb up on to the wall. There is a comfortable footpath, protected by iron railings. Below me, crowds of tourists squeeze through the gate, flowing in opposite directions, entering and leaving the city. I look to the north, with my back to the Old City, and beyond the hustle and bustle of the streets, beyond the stalls of the souvenir vendors, the busy traffic on Sultan Suleiman Street, and the small shops on the other side, I suddenly see a steep rock face with horizontal strata of white limestone. In one place, the face of the cliff turns inward, and I can see the entrance to a deep cavern. Its rounded shape shows it to be a natural cave, created by dissolved limestone. Could that be it? Is this the entrance Christ used to descend into hell? It looks exactly like Gustave Doré's engraving of the entrance to Dante's hell, except that the area around it has been built up a little since then.

But once you have passed through the natural entrance, you can see that this is not hell. You find yourself in an enormous system of straight tunnels hacked out of the rock. These are the quarries of my namesake King Solomon, or Sultan Suleiman if you prefer. Here the *meleke*, the 'kingly stone' from the Cretaceous period, was excavated

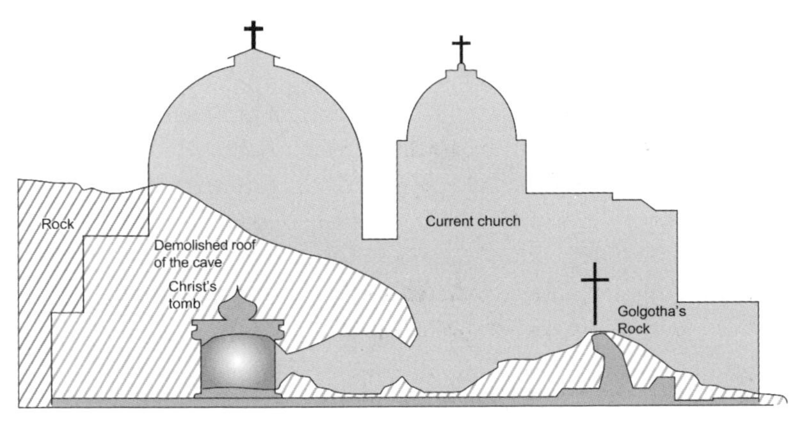

Church of the Holy Sepulchre, showing what was removed during construction.

Behind the houses lies the cliff with the entrance to Solomon's Quarries below Jerusalem, as seen from the Damascus Gate.

Gustave Doré, *The Entrance to Dante's Inferno*, 1857.

3,000 years ago to build the Temple of Solomon, and was later used for the Wailing Wall and a large part of the city centre. The tunnels extend for 400 metres, far beneath the Old City. This is the fifth quarter of Jerusalem.

The fifth quarter is also on the southwest side of the city: in limestone caves near Ketef Hinnom, on the south side of Mount Zion, tombs have been found dating back to the seventh century BCE. Two small silver amulets were found in the burial chambers bearing inscriptions to offer protection against evil: 'Yahweh is our restorer and rock.' The inscriptions are the earliest known texts from the Bible.

Ketef Hinnom is on the right side of the Valley of the Son of Hinnom, known as Gehinnom in Hebrew, Gehenna in Yiddish and Jehannah in Arabic. Gehenna, the place of eternal punishment! This is where it should be, exactly as described in Joshua 15:8. This was where King Solomon sacrificed to strange gods, and where even children were sacrificed to the idols Moloch and Baal. Yahweh warned the people of Jerusalem in the book of Jeremiah that if they did not stop worshipping idols and sacrificing children, this place 'shall no more be called Tophet ['roasting place'], nor the Valley of the Son of Hinnom, but the valley of slaughter'.

Later Gehinnom became a refuse tip where fires burned constantly, into which, according to tradition, the bodies of criminals were thrown. That is probably the reason why the Greek word 'gehenna' is translated as 'hell' in the New Testament. But there it has lost its topographical meaning and is used only allegorically.

Today Hinnom is a dry valley, a *nahal* or *wadi*, created long ago by the Hinnom river cutting through the same white kingly limestone. Further to the south, the river flowed into the Kidron and finally into the Dead Sea. There are a few more shallow natural caves in the side of the valley. It is a pleasant place to walk and there is a narrow asphalt road running through it, the pitch perhaps a small reminder of hell.

But the valley of Gehenna seems to have nothing to do with the Underworld. The ancient Israelites knew the caves around Jerusalem, excavated rock from them and buried their dead in them with silver amulets as burial gifts, but they did not associate the area with the Gehenna that Yahweh envisaged for them.

Gehenna, the 'Valley of the Son of Hinnom'.

Sulfur soap from the Dead Sea.

What hell actually was in the Bible remains unclear. In the Old Testament, the Hebrew word '*sheol*' is translated as 'hell', but also as 'grave' or 'pit'. Little more can be extracted from the texts than that it was deep under the ground and that fire burned there. There are no clues as to where *sheol* was, and it did not stink of sulfur. In the Old Testament, sulfur does not come out of the ground, but falls from the sky, like rain. That may be a reference to the eruption of Thera around 1600 BCE, or perhaps came from one of the volcanoes in the Golan Heights, though they are actually too old, or from the sulfur deposits at Be'eri on the coast, but certainly not from Gehenna or *sheol*.

Many Bible stories can be explained by geological phenomena. The Israeli geologist Y. K. Bentor, for example, attributes the destruction of the city of Hazor and the collapse of the walls of Jericho to earthquakes, the transformation of Lot's wife into a pillar of salt to partial dissolution of the salt deposits in Mount Sodom, and the burning bush to natural gas combustion. But not a word about hell and the caves.

The only text in the Bible that gives us some insight into the nature of the Underworld can be found in the Revelation of St John. It tells us of how the Devil was first cast into the abyss for 1,000 years, after which he was released. Then the beast, the false prophet, the Devil himself, followed eventually by 'the fearful, and unbelieving, and the abominable, and murderers, and whoremongers, and sorcerers, and

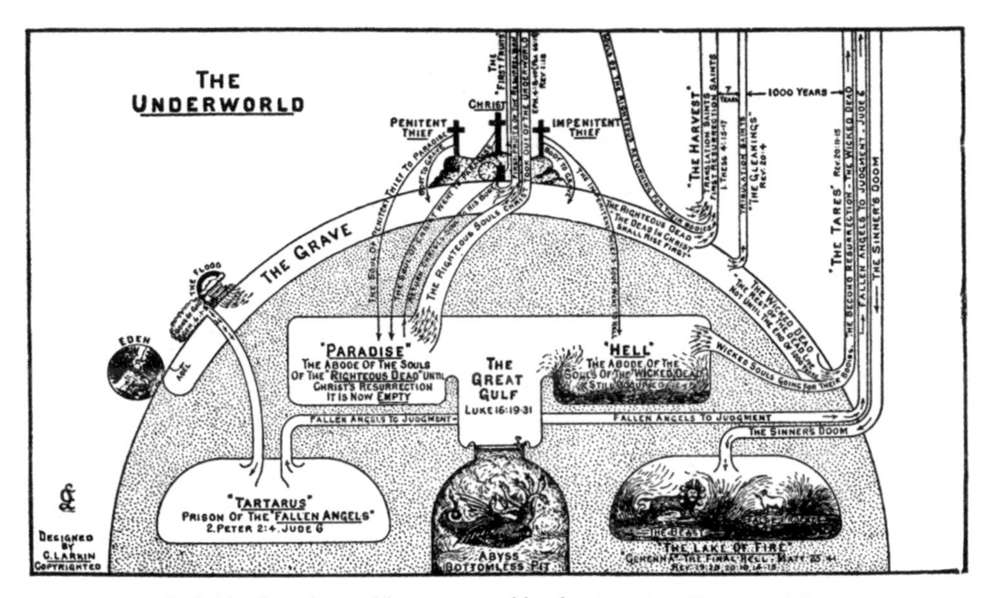

The biblical Underworld as portrayed by the American Baptist minister Clarence Larkin, 1920. There is even a small Paradise for the 'righteous' souls of the patriarchs who accompanied Christ in his ascension to Heaven.

idolaters, and all liars', was cast into 'the lake which burneth with fire and brimstone'. The abyss and the lake of fire and brimstone are therefore clearly different places, suggesting that there is some degree of spatial planning in the Underworld, though it is by no means as neatly structured as a gobstopper. And there is sulfur, though the New Testament calls it brimstone. And it does come from a lake. Finally! The Bible is like a detective story: all is revealed only at the end.

The best portrayal of the Last Judgment before Dante: a mosaic in Torcello, near Venice, *c.* 12th century.

THREE

The Wanderings of Odysseus

Midway along the journey of our life
I woke to find myself in a dark wood, for I
had wandered off from the straight path
Dante Alighieri, *Inferno*, 1:1-3

A snowy February morning in Tartu, Estonia: a city that I knew until then only from my stamp collection and from poring over old atlases and dreaming of travelling to distant lands. And now I have to be there in person; 'have to' in the pleasant sense that I am able to mix business with pleasure. I see many detached, Russian-looking, green-painted wooden houses; a market square with bumpy cobblestones; in a narrow street, the imposing neoclassical building of the Tartu Ülikool, the University of Tartu; on a wooded hill, a bombed-out church without a roof, kept warm only by the thick layer of snow on its stepped buttresses – and there he is, my hero for today: Karl Ernst von Baer. He strikes an imposing figure, sitting, almost sprawling, in his armchair, deep in thought. His right hand is supporting his head and the left is hanging resolutely over the back of the chair. The snow makes the bronze book on his lap illegible. Perhaps it is Homer.

A short distance away is the large, white-painted house where Von Baer lived for the last nine years of his life until his death in 1876. Now it is a museum. The enthusiastic young historian Erki Tammiksaar is the custodian of Von Baer's extensive scientific legacy. The study is relatively empty: there is a bust, maps of his journeys on the walls, and a drawn portrait. Von Baer has a forbidding mouth, piercing eyes, wild hair and an enormous nose. He looks intimidating; not someone you would easily dare to contradict. You see him on every Estonian two-*kroon* banknote but here, his death mask softens his features. In a glass case is displayed his book *Beiträge zur Kenntniss des Russischen*

Reiches und der angrenzenden Länder Asiens, published in 1872 in St Petersburg.

Erki entices me into a side room that contains all the publications by Von Baer he has been able to collect. The man wrote profusely on the most widely varied subjects and in the most unlikely journals and this is the only place where they have all been brought together. Erki has spent months tracking them down in libraries in St Petersburg and Giessen. There is no decent biography of Von Baer, only a Russian one that is so full of errors and plagiarized passages that Erki cannot disguise a snigger when he tells me about it. Amazing that renowned scientific publishers Nauka allowed it to be published, he says. He is making it his life's work to write a good biography, with the unwavering dedication that only a true admirer can marshal.

Karl Ernst von Baer (1792–1876) was a scientific prodigy. He was born in Estonia, studied medicine and was the founder of modern embryology: he discovered the female ovum in mammals and humans, and corresponded on his discovery with other scientists, including Charles Darwin. Before then, no one had a clear idea of how reproduction worked.

I first became acquainted with his work during a visit to the Caspian Sea. Near Astrakhan, in the Volga delta, there is an area of dead straight sand hills tens of kilometres long and around twenty metres high. It looks as though an enormous harrow has passed over the landscape. Astrakhan itself lies on one of these hills. Von Baer was the first to describe the hills in 1854 in his extensive Caspian diaries: he was not only a physician, but also an explorer and geologist, not to mention an ethnographer, anthropologist and ichthyologist. By then he had been appointed head of the Anthropological Museum in St Petersburg, and was a member of the St Petersburg Academy of Sciences. His Caspian expeditions were part of his countless scientific travels in Russia and the surrounding countries, which he recorded in the book that I saw in the display case in Tartu.

The elongated hills in the Volga delta are still known as the *Bèrovskiё bugry*, the Baer hills. Von Baer claimed that they were dunes and to this day, 150 years later, there is heated discussion among Russian geologists as to whether that is the case. I think he was right. For that reason alone, I wanted to travel to Tartu: to track down Von Baer's original manuscripts.

The second reason was Homer, and his account of Odysseus descending into the netherworld. The goddess Circe demanded that

Odysseus pay a visit to the blind seer Teiresias in the Underworld before leaving her island, Aeaea, to return home. In the *Iliad* Homer describes the Underworld as he saw it (VIII:13–16). At the start of a gathering of the gods on Mount Olympus, Zeus gives a thunderous speech in which he threatens anyone who dares to help the Trojans or the Danaans in the Trojan War: 'I will hurl him down into dark Tartarus far into the deepest pit under the Earth, where the gates are iron and the floor bronze, as far beneath Hades as heaven is high above the Earth.' Hades is already under the ground, and Tartarus is apparently even deeper. As in Revelations, we see signs of a layered structure in the Underworld. The gobstopper is getting closer.

But Odysseus does nothing with this information. His famous journey to the Underworld in the eleventh book of the *Odyssey* is actually an anticlimax. In fact, he never actually goes there at all. Circe does tell him how to get to the mildewed kingdom of Hades. First he has to cross the raging Oceanus, and then beach his ship on the shore near to the groves of Persephone. From there, he has to go to a rock where the rivers Pyriphlegethon and Cocytus, a branch of the river Styx, run together and flow, roaring violently, into Acheron. There he has to dig a trench a cubit long and a cubit wide and pour into it a libation to the dead consisting of wine, honey, milk, water and barley meal.

Odysseus takes her advice, arrives at the designated place – in the misty kingdom of the Cimmerians at the very end of the Oceanus – digs a trench with his sword, as Circe had instructed, and makes the libation. The ghosts rise up from the Underworld, emerge from the trench and tell their stories, one after the other. So Odysseus does not have to go down there himself, and is spared its horrors. That's a pity: I would have liked to hear what it was like down there.

This is, however, the first time that the entrance to the Underworld is identified geographically, and I want to know where it is. Nineteenth-century writers like Johann-Heinrich Voss, the cartographers of antiquity Karl Spruner, Heinrich von Merz and Theodor Menke, and even the politician William Gladstone, have sought it with the tenacity of Atlantis-seekers. According to the cartographers of the time, Oceanus encircled the whole known Earth, and some took this literally, searching for the place where Odysseus beached his ship in the Atlantic Ocean, past the Strait of Gibraltar, which was the Scylla and Charybdis of the time. It was always misty in the land of the Cimmerians, Homer wrote, and that strengthened their conviction that the entrance lay somewhere in the north.

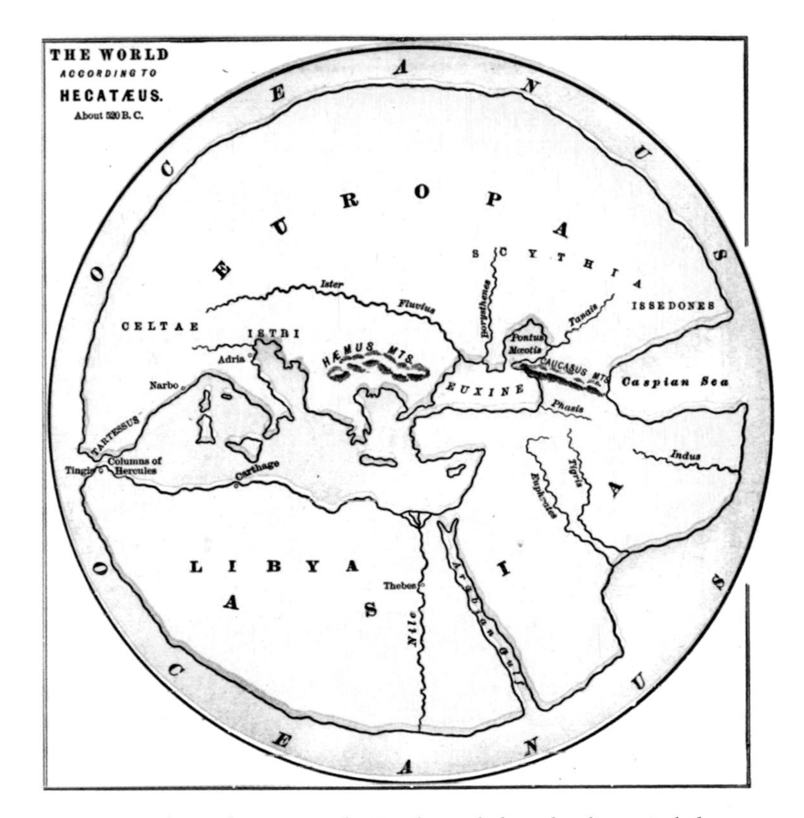

In classical antiquity, the Earth was believed to be encircled by the great river Oceanus.

During his travels through the Russian Empire, Karl Ernst von Baer had explored the northern coast of the Black Sea in great detail, and had become convinced that Odysseus had dug his trench to the Underworld there. He even wrote a book about it, which was published posthumously in Braunschweig in 1878 with the title *Über die Homerischen Lokalitäten in der Odyssee*. Shortly before he reached Circe's island, Odysseus had paid a visit to the Laestrygonians.

> When we reached the harbour we found it land-locked under steep cliff, with a narrow entrance between two headlands. My captains took all their ships inside, and made them fast close to one another, for there was never so much as a breath of wind inside, but it was always dead calm. (x:87–94)

The visit didn't go well. The Laestrygonians proved to be man-eating giants and speared several of Odysseus' companions for their dinner.

But that was not what interested Von Baer: 'The description of the bay of the Laestrygonians so closely resembles that of the bay of Balaklava that it cannot be a coincidence.' Balaklava is an old city in the Crimea which has now been incorporated into the nearby port of Sevastopol, the home of the Russian Black Sea fleet. Its bay is a striking S-shape. Von Baer writes that he went there twice and both times the water was dead calm, despite the choppy waves on the open sea outside the bay. Homer could not have made that up, he adds, as such bays do not exist along the Greek coast (this is not true, as we shall see later). For Von Baer, this is the ultimate proof that the whole of Homer's story must have taken place in the Black Sea. He even succeeds in locating Persephone's groves: they are woods of poplar near the mouth of the Kuban River in the Sea of Azov.

And what is more, the Cimmerians lived there, as Herodotus wrote in his History (4,12): 'Scythia still retains traces of the Cimmerians;

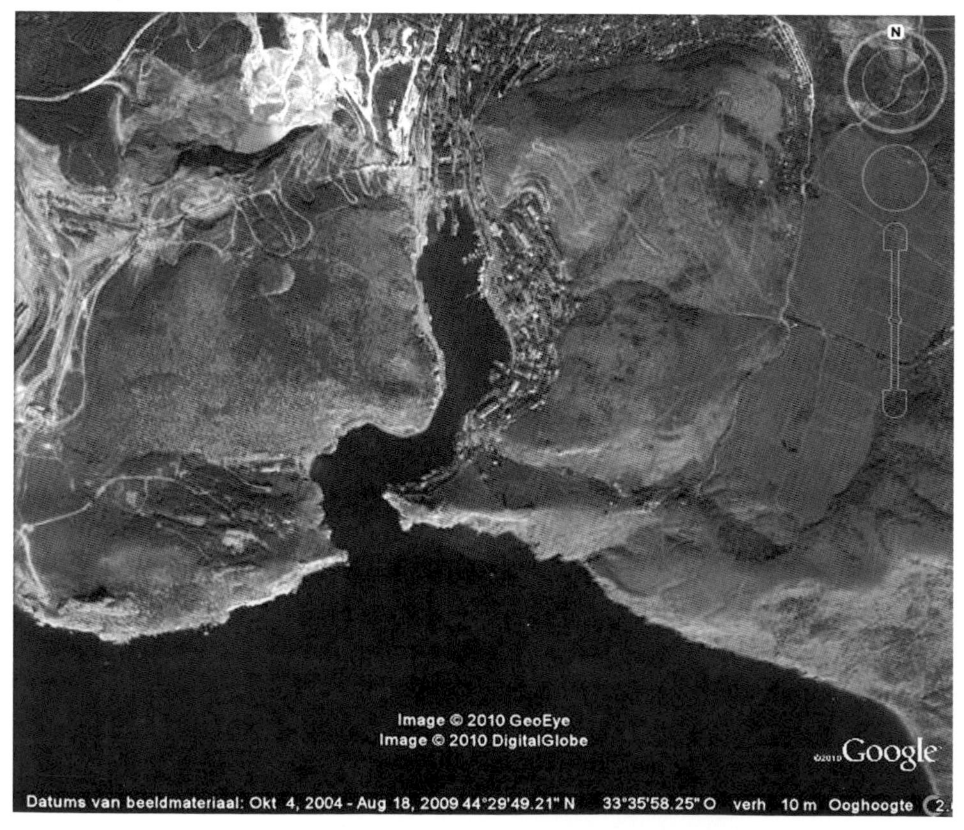

The Bay of Balaklava-Sevastopol – according to Von Baer the bay of the Laestrygonians.

there are Cimmerian castles, and a Cimmerian ferry, also a tract called Cimmeria, and a Cimmerian Bosphorus.' The latter is the classical name for the narrow Strait of Kerch between the Black Sea and the Sea of Azov or, in landlubbers' language, between the Crimea and the Caucasus. Even the name 'Crimea' may be a reference to the Cimmerians, coming from the Akkadian *Gimiru*, meaning 'those who migrate back and forth'. The Greek colonies of Gorgippia, Phanagoria, Pantikapaion, Nymphaion, Kimmerikon and Theodosia were clustered around the strait 'like frogs around a pond'.

The Styx, Von Baer argues, is known as a black river, and Pyriphlegethon means 'flaming like fire'. On both sides of the Strait of Kerch there are countless mud volcanoes, especially on the Russian side near Taman. These small, simmering cone volcanoes do not expel lava or volcanic ash, but liquid mud that plops up in bubbles or flows in slow streams. The mud often contains oil, which makes it look black, or gases that can catch fire, as we shall see. 'If you look down into the cones, you see a black, bottomless pit and it takes little to imagine spirits or the dead rising from these hollow shafts.' Von Baer then claims with great aplomb that 'there can be little doubt that this region inspired the images that the ancient Greeks used to portray Hades or Aides.' In 1877, in the cultural journal *De Gids*, classicist Samuel A. Naber repeated them almost verbatim: 'We can see that this region accommodates all the images that belong to the Greek Underworld: muddy rivers, some flowing with black water, others with currents of flame.'

But other classicists in Von Baer's time called his ideas pure fantasy, and there are a number of dubious aspects to his reconstruction. Circe's island, Aeaea, is not an island at all, but refers to the Mingrelian coast of Georgia; Calypso's island, Ogygia, lies in the Alps, something which Von Baer has to go into all kinds of contortions to explain. In the *St Petersburger Zeitung* of 26 July 1875, he published a sneering reply to these critics, claiming that all the so-called Hellenists sat in their studies like badgers in their sets, and if anyone got too close, they bit randomly in all directions, just like badgers do. They should all get out there and see for themselves.

It is September 2007. By coincidence, I 'have to' go the Black Sea, to the land of the Cimmerians. I am travelling from Gelendzhik on the Russian Black Sea coast to Odessa with a motley company of Russians, Ukrainians, Canadians, Iranians, Moldovans, Turks, Americans, Britons, Germans and Azerbaijanis. We have just discussed the history of the sea level of the Black Sea at a congress in Gelendzhik. Some claim that it has risen and fallen quite spectacularly in the past 1,000 years, while others contest that. Now we can see the evidence. We see in the steep cliffs on the southern side of the Taman peninsula how high the sea level was in the previous warm period. We see the long spit of Tuzla Island in the Strait of Kerch, which continually shifts, causing recurring border conflicts between Russia and Ukraine. We find the indications of recent changes in the sea level less than convincing.

There is also a small museum dedicated to the writer Michail Lermontov. 'Taman is the nastiest little hole of all the seaports of Russia', he writes in the gloomy story 'Taman', part of his masterpiece *A Hero of Our Time* (1839). The hero is the travelling officer Grigoriy Pechorin, who is robbed in the mist by a blind man. And so Homer's blind seer Teiresias and the misty land of the Cimmerians return to world literature in different guises. Even the name 'Taman' means mist.

We also see 6,000-year-old solid dolmens from the Stone Age, kurgans, burial mounds with ingenious domed roofs, and the ruins of the Greek colonies of Gorgippia (now Anapa), Phanagoria and Pantikapaion (now Kerch). The museums in the towns exhibit Greek statues and amphorae, Persian and Jewish-Khazar gravestones, and magnificent golden Scythian jewellery and Khazar necklaces of turquoise and agate.

But there is no trace of the Cimmerians. Not a single archaeologist has been able to prove that they were ever here, despite the fact that Homer and especially Herodotus wrote so emphatically about them. I ask our guide Viktor Zinko about the Cimmerians. 'We know now that they lived on the south side of the Black Sea, not the north', he tells me.

Taman's mud volcanoes are still there and are active, but Homer doesn't actually mention them specifically. Homer may in fact never have been here, and only knew the places he mentioned through hearsay. Perhaps he didn't want to make a route description at all, for Odysseus, for Von Baer or for me, just tell a good story. But surely Herodotus would not have lied? The puzzle has still not been solved.

On 11 November 2007, two months after my departure, all hell breaks loose in the Strait of Kerch. During a storm, the oil tanker *Volgoneft-139* breaks in two and discharges 4,000 tons of oil into the sea, polluting 30 kilometres of coastline and killing 30,000 birds. Four other cargo ships sink, one carrying scrap iron and the other three with sulfur on board. Some 7,000 tons of sulfur disappear into the

The Strait and Bay of Kerch, seen from the ruins of the Greek settlement of Pantikapaion. 'Stability and prosperity' is the message on the election billboards of the pro-Russian Viktor Yanukovych's Party of the Regions, and the Russian flag flies here on Ukrainian territory.

sea. It is as though Von Baer is trying to tell us from the hereafter that he was right by depositing sulfur at the place where he was certain the entrance to hell must have been.

FOUR

The Entrance to Hell

> Abandon every hope, all you who enter.
> Dante Alighieri, *Inferno*, iii:9

It must be here. I park my rented car on the side of the road and start to walk around the lake, Lago Averno, near Naples. Here, in the 'foul jaws of stinking Avernus', Virgil's hero Aeneas descended to the Underworld to ask the advice of his dead father, Anchises.

> There was a deep stony cave, huge and gaping wide,
> sheltered by a dark lake and shadowy woods,
> over which nothing could extend its wings in safe flight,
> since such a breath flowed from those black jaws,
> and was carried to the over-arching sky, that the Greeks
> called it by the name Aornos, that is Avernus, or the
> Bird-less. (vi:237–42)54.504 mm

Virgil's description of Lago Averno is the most concrete reference to the entrance to the subterranean world in the whole body of classical literature. Yet strangely enough, there is nothing sinister to be seen there. It is a friendly, round lake, lined with steep slopes covered with vineyards and fruit trees. As Virgil says, Avernus (*Aornos* in Greek) means 'birdless place', but ducks and coots dive into the clear water between the high reeds, and 'extend their wings in safe flight'. A man on a light-blue Vespa, his small son on the back, comes popping past me on the narrow footpath. At Bar Caronte, there is no ferry across to the darkness on the other side, but a cheerful group of people enjoying their lunch. The mineral water may be called *Lete*, after the underground river of forgetfulness, but its quality is guaranteed by the Sapienza University in Rome.

What did Virgil have against this place, which is more like Paradise than hell? Where are the cave, the grove and the poisonous vapours?

Or is it all just poetic licence? Did he use the name simply because it sounded good and fitted nicely into his metre? Is it a fable that birds that tried to fly over the lake dropped dead into the water? Not at all: around 55 BCE Lucretius had already noted in his didactic poem *De rerum natura*:

> First of all, as to the name Avernian, by which they are called, it has been given to them from their real nature, because they are noxious to all birds; for when they have arrived in flight just opposite those spots, they forget to row with their wings, they drop their sails and fall with soft neck outstretched headlong to the earth, if so be that the nature of the ground admit of that, or into the water, if so be that a lake of Avernus spreads below. There is such a spot at Cumae, where the mountains are charged with acrid sulfur, and smoke enriched with hot springs. (VI:740–48)

Virgil was familiar with Lago Averno at first-hand, since, in the years that he wrote the *Aeneid* (between 29 and 19 BCE), he lived largely in Naples.

The Greek geographer Strabo, a contemporary of Virgil, describes the current location of Lago Averno very precisely in *Geographica,* his comprehensive encyclopaedia of the world, and refers to it as the entrance to the Underworld. He also uses the Greek name Aornos, and tells of how birds fall into the water because of the strong vapours arising from it. Although Strabo wrote *Geographica* after Lucretius' *De rerum natura* and Virgil's *Aeneid*, he does not refer to them, but almost exclusively to Greek sources. According to his biographer Anna Maria Biraschi, he had little respect for Latin writers. Many independent sources therefore clearly saw Lago Averno as a dangerous lake over which birds should not fly.

Lago Averno is not just a round lake, but a crater lake. It was once the site of a volcanic eruption, and therefore it is not illogical to expect to see vapours arising from it. I walk along the water's edge to try and spot rising bubbles of gas, just as I do with my students every year along the shores of the Laacher See, a large crater lake in the Eifel region of Germany. There gas bubbles rise to the surface day and night, throughout the year, even through the ice in the winter. If you stand still and listen, you can hear a soft, friendly sound like a simmering saucepan, an unremitting emission of carbon dioxide, 2,500 tons

Pietro Fabris, *Lago Averno*, 1776, colour engraving. Lago di Lucrino is just visible between the lake and the sea.

a year since the last eruption 13,000 years ago, to remind us that the Earth is still very busy down there beneath our feet. I feel very insignificant when I think that my great-great-grandparents could have heard it and that my great-grandchildren will be able to, too. A man is nothing more than a single air bubble, breaking away from the muddy ground, rising to the surface, forming a short-lived, beautiful dome a centimetre across, and then popping and releasing its gas into space.

But at Lago Averno none of this is to be seen; just ripples on the water, reeds, ducks and duckweed. On the eastern side of the lake there is a Roman ruin of thin, flat bricks rising high into the air. The roof has collapsed, but the high windows show that the building had at least two floors. According to the map, it is the 'Tempio di Apollo', but renowned Neapolitan archaeologist Amedeo Maiuri, who has also made spectacular excavations in Pompeii and Herculaneum, claims it is not a temple but the remains of a gigantic complex of thermal baths from the Roman period. The central hexagonal hall was almost as large as the Pantheon in Rome. But where were the hot springs? They would be evidence of volcanic vapours in antiquity. Even the poet Boccaccio wrote in 1337 – thus after Dante's *Divine Comedy* – about the warm springs of Averno. But I can't find them.

I have nearly walked around the entire lake, when I suddenly see a sign saying '*Grotta della Sibilla, a 50 m*', with an arrow. So there is a cave! But it turns out not be a real cave. Under the shade cast by leafy sweet chestnuts, I follow a cool gully dug out of yellow tuff, which leads to a dark tunnel closed off after a few metres by a red gate. Is the gate there to stop people descending into the Underworld on their own initiative? I squint through the gate into the darkness and then take a flash photo. I can see on the screen that the floor of the tunnel is 20 centimetres underwater. Who would dig a tunnel under water? Or is this the Styx? If it is, it's something of a disappointment. You wouldn't need a ferryman; if the gate were not locked, you could just wade through it. But where else does the tunnel lead to? And there aren't any vapours. There is nothing volcanic about it, and it all seems to be manmade. Am I in the right place?

Yet I start to regain my confidence in Virgil. After all, Sibyl was the prophetess who told Aeneas how he could go down to the Underworld! What does Virgil say about her cave?

> The vast flank of the Cumaean cliff is pitted with caves,
> from which a hundred wide tunnels, a hundred mouths lead,
> from which as many voices rush: the Sibyl's replies. (*Aeneid*, VI:42–5)

The problem is, Cumae is 3 kilometres away from here on the coast, on the other side of the 100-metre-high rim of the crater! How can this flooded little tunnel be the entrance to Sybil's cave? Or does it run all the way through to Cumae? Nor does it have 'a hundred wide mouths', but just the one, its lips pursed tightly together. Is Sybil's cave not the entrance to the Underworld at all, but something else? Surely it's very unlikely that the oracle could sit here foreseeing the future while the birds were dropping, stupefied, out of the sky? And yet the sign pointing to the Grotta della Sibilla must be here for a reason. Perhaps there are two caves near Lago Averno, one leading to the Underworld and the other to Sybil. I have to go to Cumae!

It is late afternoon when I park under the black plane trees. In front of me, the vertical walls of a steep tuff plateau rise up from the narrow coastal plain. An ideal spot for a fort. The footpath leads to an enormous arch, hacked out of the yellow tuff in a spur of the rock face. It's the same stone as in the tunnel in Averno: fine, yellow ash containing large chunks of lava and hardly layered. The rock glows slightly in the low evening sun. To the right of the western exit, metres-deep cisterns have been hacked out of the tuff, and are still partially lined with Roman bricks. On the top of the plateau are the ruins of the

Temple of Apollo, where Sybil was a priestess. But I have to go to the left, as there lies the Antro della Sibilla, Sybil's cavern, as it is known here. This must be the real thing – and not only because it is in Cumae. It is a passage more than 100 metres long, hacked out of the tuff, parallel to the western face of the tuff plateau with, at regular intervals, side passages to the outside through which the bright sunlight shines: Virgil's hundred mouths. It is a fascinating *chiaroscuro* if you look down the passage, an effect that is enhanced by the trapezoid upper half of the tunnel, which is almost 5 metres high.

The centuries-old voice of tragic Sybil must have echoed through these passages. She was tragic because, as Ovid relates in his *Metamorphoses*, she asked Apollo for as many years as there are grains of sand in a sand heap, but failed to ask for eternal youth. Apollo was prepared to give her eternal youth, but only in exchange for her chastity, which she refused to surrender. She told Aeneas that she had already lived for 700 years, that she had another 300 to go, and that her body would shrivel to a tiny heap until only her voice remained. It sounds as though she regretted her earlier prudery.

She had been alive for 700 years! Exactly from the time when the Greeks founded Cumae in the eighth century BCE! Was that intentional or coincidence? The truth or poetic licence? The first reference to the Antro itself dates from the third century BCE, though there is some disagreement about that. The Greek poet Lycophron of Chalcis wrote in his *Alexandra* that 'the grim dwelling of the maiden Sibylla, roofed by the cavernous pit that shelters her' was at the hill of Zosterius (Apollo). But he did not say that the cave is in Cumae. Archaeologists think that, because of the shape of the passages, the Antro must date from the sixth or fifth century BCE. Canadian classicist Raymond J. Clark believes that it was the work of the sixth-century Greek tyrant Aristodemus. One thing is certain: the Antro della Sibilla was there when Virgil wrote about it.

Surely, once you've seen this incredible cave, you forget the small tunnel at Lago Averno? So why would everyone since the Renaissance believe that the tunnel was the Grotta della Sibilla? The reason is that the Antro della Sibilla in Cumae was only rediscovered in 1932, excavated by the same Amedeo Maiuri whom we have previously encountered. Before then, no one had noticed the Antro because, as Maiuri writes in his small book *I Campi Flegrei*, the entrance and the first 25 metres had been destroyed by old quarry works and were buried under rubble and waste. The entrance itself was concealed by

Cave of the Antro della Sibilla, Cumae.

a simple field oven, while the whole passage was blocked with loose stones and earth that had fallen in through the side openings. Only the extreme southern opening was partially free and that was used by a farmer to store his wine. It is one of the most exceptional, suggestive and inspiring monuments not only in Italy, but also in the whole of Mediterranean civilization, writes Maiuri, not without some pride. But it too emits no vapours and was hewn by human hand. The Antro della Sibilla is no more the entrance to the Underworld than is the tunnel at Lago Averno.

So what *is* the tunnel at Averno? Maiuri was permitted to enter it in the 1930s. For him the red gate did open. It is a dead straight passage, 200 metres long, and it leads not to Cumae, but to the west, cutting through the steep tuff ridge to the next lake, Lago Lucrino. Lucrino is not a crater lake, but a coastal lagoon, separated from the sea by a beach ridge. We also know this from the Greek geographer Strabo. The tunnel was part of a gigantic military port complex that the Roman ruler Octavian, later emperor Augustus, commissioned his brilliant general Agrippa to build in 37 BCE during the civil war that followed the assassination of Julius Caesar. The largest part of the port, the Portus Julius, now lies under 10 to 12 metres of water in the blue bay of Pozzuoli. In clear weather the ruins can be seen from the air. Agrippa breached the beach ridge to make Lago Lucrino part of the port. He also had a canal dug from Lago Lucrino to Lago Averno, through the only low point in the steep crater wall, through which the access road now passes. The protected round crater lake was invisible from the sea and therefore an ideal hiding place for Agrippa's fleet. He chopped down the forests around the lake and cultivated the steep slopes of the crater. There were buildings all along the shores, the ruins of which can still be found. The tunnel was probably used to move troops in secret. There is a second tunnel, on the western side of the lake, which does run to Cumae and is known as the Grotta di Cocceio, most probably after its constructor. This second tunnel is also no longer accessible. In the Second World War it was used to store explosives, which blew up, causing the tunnel to collapse. What was made for one war was destroyed by another. But none of this had anything to do with Sybil.

Agrippa's intervention is confirmed from an unexpected quarter. German palaeontologist Francisco Welter-Schultes of the University of Göttingen has recently taken bore samples from the sediment on the bed of Lago Averno, and discovered a strong increase in seashells exactly in the year 37 BCE. This was the year when the canal was reported to have been dug – and Averno would have been connected to the sea. Above and below the layer of seashells, freshwater shells dominate. Pollen grains in the same bore sample provide evidence of deforestation around the same time: before 37 BCE, they are largely from holm oaks that grew on the rim of the crater. After that, the picture changes abruptly and agricultural crops dominate. The influence of the sea is estimated to have lasted around 70 years, after which the canal is assumed to have fallen into disuse.

Did Virgil then not know what had happened to the lake? Of course he did. After all, he wrote the *Aeneid* between 29 and 19 BCE, after Agrippa's construction works. And he was living in Naples at that time. He even describes the link to the sea in an earlier work, the *Georgics*, which he must have started around 37 BCE:

> Shall I recall the harbours, and the barrier across the Lucrine,
> and the angry ocean sounding, far off, in mighty anger
> where the Julian waves are repulsed
> and the Tyrrhenian tide pours into the straits of Avernus?
> (II:161–4)

So what remains of the threatening birdless lake, from which toxic vapours rise into the air? In Virgil's time it was a saltwater sea inlet, where ships were moored and wealthy Romans bathed in the largest thermal complex in the region! Is it all poetic licence after all? Am I searching in vain here for the entrance to hell? But Virgil may also have seen it before Agrippa made his rigorous changes. My main source of evidence is Lucretius, who described the vaporous pit around 55 BCE. Did the pit fill with water when Agrippa had the canal dug to the sea?

At the age of seventeen, I travelled with my school mates through Italy, an excursion organized by my high school: Florence, Rome, Naples, Ravenna, Venice, all by train. It was actually a modern version of the Grand Tour that artists used to make through Italy in the seventeenth and eighteenth centuries to acquire inspiration. Only now, during my own descent into hell, do I realise that I must have seen Lago Averno then, too, from the plateau at Cumae – at least, there was a crude sketch of the Antro della Sibilla in the letters I sent home.

A highlight of the trip was the Solfatara crater, not very far from Lago Averno and Pozzuoli, which does have a 'stinking pit'. It is a similar

Deposits of sulfur and arsenic sulfides in the Bocca Grande, Solfatara.

crater to Lago Averno, round and with steep sides, and about the same size. There are not many trees, only a small wood on the western side of the crater, where there is a campsite (if you ever want to camp in a crater, this is the place), but otherwise it is a wilderness in the middle of the chaotic and fully built-up suburbs of Naples. Plants refuse to grow there because, biologists say, the ground is too hot. Local ornithologist Mariella Abatino writes that some birds can be found in the wood to the west, but there are few to be seen flying over the central part of the volcano. So the Solfatara is a little *aornos*. There must have been a lake there, too, in former times, but it probably evaporated. All that is left is the Fangaia, a bubbling mud pool in the middle of the crater. I place my photos of the Fangaia next to each other, the first from 1964 and the other from 2009, and see little difference. The mud is still bubbling, and little has changed in the 45 years of a human life.

The most spectacular part of the Solfatara is the Bocca Grande, a fumarole where steam containing hydrogen sulfide is expelled into the air under great pressure and at a temperature of 160°C. Hydrogen sulfide oxidizes in the heat and deposits bright yellow sulfur crusts on the edges of the Bocca Grande. Orange-red realgar (arsenic sulfide) and red cinnabar (mercury sulfide) are also deposited, producing an even livelier palette of colours.

In the whole crater, the smell of sulfur is pleasantly stimulating and, since Roman times, this has been a place where people come to heal their ailments. Strabo himself calls it the Agora of Hephaistos, or in Latin Forum Vulcani, adding that it stinks horribly. The thirteenth-century Neapolitan text *Trattato dei bagni*, based on *De balneis Puteolanis* by Pietro da Eboli, notes that a bath in the Solfatara

> settles the nerves, alleviates scabies and strengthens enfeebled limbs. It also helps against headache and stomach-ache, protects against fractures and improves the eyesight. It will fight a fever, especially if you purge yourself before taking a bath. You should not be afraid of the stench as this drives the sickness from your body.

In the Middle Ages the infirm could have a bottle filled with gas from the Solfatara delivered to their homes.

There are two sudatoria, sweat rooms, dug into the crater wall, shallow tunnels lined with bricks, exuding hot vapours. There is a sign saying '*Sauna vietata*', saunas forbidden. I remember that the rector of our school, the classicist Dr H. W. Fortgens, went into one of the steaming holes, a knotted handkerchief on his head, to take the vapours in front of his pupils. But he did not last out very long in his sloppy grey suit. The hot vapours do indeed arouse an irresistible desire to remove your clothes.

Medicinal miracles are by no means a thing of the past. According to the Italian newspaper *La Repubblica*, since a team of American-Neapolitan scientists led by Louis Ignarro published in the prestigious open-source journal *Proceedings of the National Academy of Sciences* in 2009 that hydrogen sulfide also helps with erection problems, there has been an incessant flow of tourists.

But in the Solfatara the gas that bubbles up out of the Fangaia and is expelled into the air in the Bocca Grande comprises 99.2 per cent CO_2 (carbon dioxide) and only 0.5 per cent H_2S (hydrogen sulfide), not to mention a lot of steam. A small amount of hydrogen sulfide is enough to stimulate the nose. There are large round bottles with hoses stuck into the ground to measure the outflow and composition of the gases.

Why did Virgil not locate his entrance to the Underworld in the Solfatara? After all, it has everything: sulfur, the stench, steaming, birdless pits. Perhaps it was because it was a well-known spa, which the Romans sooner associated with healing and pleasure than with death. And perhaps because there was a steaming pit at Lago Averno. Virgil doesn't talk about the 'foul jaws of stinking Avernus' without reason. Carbon dioxide doesn't smell; it was sulfur that came up out of hell.

An old memory rises to the surface. My first geology book was *Algemeene Geologie* by B. G. Escher, the 'fourth, much extended edition' of 1934. It was a solidly bound book with a brown linen cover, published by the Wereldbibliotheek. It is still a wonderful book, full of drawings by Escher himself. He had the same talent for drawing as his half-brother, Maurits Escher, and shared his fascination with symmetry, writing a textbook on crystals and minerals. I received *Algemeene Geologie* on my fifteenth birthday as a present from my favourite uncle, Lipke Holthuis, curator of crustaceans at the Natural History Museum in Leiden (now Naturalis). He collected crabs, lobsters, prawns and woodlice for the museum from all over the world. I would have loved to have gone with him. He had attended lectures by Escher, and nurtured my embryonic geological interest by bringing back stones from far-off places, and by giving me this book. It was a decisive moment in my choice of studies. 'Mofettes, very weak indications of dying volcanic activity, are places where toxic carbon dioxide (CO_2) escapes from the ground. They are often located near death valleys (Tangkuban Perahu on Java, the Dog Cave near Naples, the carbon dioxide cave in Royat, Auvergne), etc.', I read in the book. Death valleys! The Dog Cave near Naples! Why was the cave called that? Even then, I wanted to know more.

I can go no further without referring to Athanasius Kircher, *uomo universale*, polymath from the seventeenth century. His biographer Paula Findlen called him 'the last man who knew everything', though French writer Jean-Marie Blas de Roblès was slightly less flattering in his novel *Where Tigers Are at Home*, noting rather maliciously that Kircher was 'a man who associated with people like Leibniz, Galileo and Huygens, and who was much more famous than them, and yet was completely wrong about everything'. We shall meet Kircher again later in this book.

The Dog Cave crops up in Kircher's *Mundus Subterraneus*, written in 1664. It is a hefty tome with splendid engravings and offers a delightful glimpse of the scientific world of 400 years ago. Kircher describes his visit in 1638 to the Cave of Charon, Grotta de Cani, or the Cave of Dogs, 'and the same deadly force and property which causes those who enter to suffocate and die'.

> This Cave lies close to a Lake they call Agnano, which is round in shape and 500 paces in Diameter. The water is clear and quite cold, and is replenished by mineral Springs. On the shore there is a Farm cottage, where the owner breeds a large number of Dogs to use for tests in the cave. As soon as we arrived at the spot, he took a Dog and tied it to a long stick, whereupon a man who knew about these things pushed the beast into the deadly Flue of the Cave. When the Dog entered the Flue he could not bear the acidity of the toxic vapours that arose from it, and appeared to suffocate and be completely unable to move. The Dog was pulled out of the Hole and submerged in the waters of the Lake and, after a short time, as though he had been roused from a deep sleep, he started to walk again. After he had been refreshed with a little food, he was returned to his Master.

Kircher wonders what causes the asphyxiation: 'if the infected vapours have their origins in Ochre, Sandrak, Orpiment and Arsenic, then

Saving a dog from the Dog Cave by immersing it in the Lago di Agnano. Engraving by Georg Hoefnagel, detail from a diptych, drawn *c.* 1555 during a journey to Italy with the cartographer Abraham Ortelius. From the atlas *Civitates orbis terrarum* by Georg Braun and Franz Hogenberg (1575), a popular supplement to the famous atlas *Theatrum orbis terrarum* (1570) by Ortelius himself. The engraving was later copied in countless books.

they caused without any shadow of a doubt a stiffening of the heart'. Sandrak and orpiment are yellow and red arsenic compounds used at that time for dyeing. Blaming arsenic is not so ludicrous since, as we have seen, arsenic crusts form around the Bocca Grande in the Solfatara. Carbon dioxide was still hardly known, being discovered around the same time by the Flemish scientist Jean Baptista van Helmont (1579–1644).

Kircher cites scientific experiments in the cave from letters he received from the Spanish-Bohemian scholar Juan Caramuel y Lobkowitz, bishop of Vigevano, near Milan. Caramuel and his companions observed that a torch continued to burn in the upper part of the cave, which was filled with cold air, explaining why it is possible to walk upright in the cave without problems. In the lower part of the cave, however, the torch went out. We know now that this is because of carbon dioxide, which is heavier than air.

SEVENTH OBSERVATION. We then let a Dog into the Cave. As long as he was able to keep his head above the CD line he stayed calm and silent, giving no indication at all that he had any difficulties breathing the air. However, when he was forced to dip his head below the line, he did everything in his power to wrestle himself loose, and escaped twice from the man who had driven him into the Cave. But once they had placed a stick around his neck, he lay still and not breathing, as though in a faint. You would think that he was dead. Then he was pulled out and submerged in the Lake, upon which he showed some signs of life but still did not stand up. He was pulled back onto dry land so that he did not drown and water was poured over him. He

The experiments by Caramuel y Lobkowitz and his companions in the Cave of Dogs. D–C is the dividing line between oxygen-rich and oxygenless air; note the torch in plates 2 and 3. Engraving from Athanasius Kircher, *Mundus subterraneus* (1664–5).

> revived, stood up, looked at us all and – for fear that he would be pushed back into the cave – walked off towards Naples.

The same fate befell a frog. Samples were taken from the soil, and the Marquis de Arenis, their Neapolitan host who also took part in the experiments, wanted to examine the air below the line, 'or how much heavier the vapour was than normal air; but the servants (as is usually the case) had not brought the Tools he had ordered'.

There are countless legends and tales about the Grotta del Cane. Pliny the Elder mentioned caves of Charon near Pozzuoli, holes that emitted deadly fumes (*Natural History* II, 95). Seneca (*Naturales Quaestiones*, VI, 28) also says that lethal vapours rise from cracks in the ground in various places in Italy, causing birds in full flight to fall to the ground and swell up as though they have been strangled. William Hamilton and later authors believed that these were specific references to the Dog Cave, but they could just as easily have referred to the Solfatara, or to the 'foul jaws of stinking Avernus', if they ever existed.

Pedro de Toledo, sixteenth-century Spanish ruler of Naples, is alleged to have imprisoned two slaves in the cave to punish them. Of course, they did not survive. Two farmers are reputed to have gone to the cave to sleep and never woke up. Besides dogs, the cave's animal victims also included chickens and roosters. De Ferrari writes in his travel guide in 1826, presumably to encourage adventurous travellers, that a rooster that had stuck its head into the cave immediately began to vomit and died on the spot. The Dog Cave, too, is *aornos*, birdless. It was a popular feature of the Grand Tour: Agricola, Spallanzani, de La Condamine, Goethe, Dumas, William Hamilton and Breislak all paid a visit. Mark Twain wrote in *The Innocents Abroad* (1869):

> Every body has written about the Grotto del Cane and its poisonous vapors, from Pliny down to Smith, and every tourist has held a dog over its floor by the legs to test the capacities of the place. The dog dies in a minute and a half – a chicken instantly. As a general thing, strangers who crawl in there to sleep do not get up until they are called. And then they don't either. The stranger that ventures to sleep there takes a permanent contract. I longed to see this grotto. I resolved to take a dog and hold him myself; suffocate him a little and time him; suffocate him some more and then finish him. We reached the grotto at about three in the afternoon, and proceeded at once

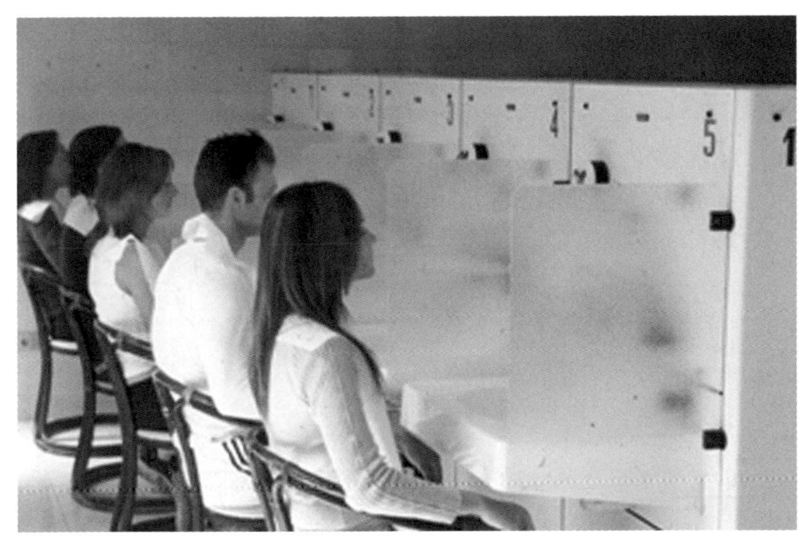

Terme di Agnano spa inhalation. The healing vapours are sprayed into the mouths of the visitors to the baths. A dog would take to his heels.

> to make the experiments. But now, an important difficulty presented itself. We had no dog.

The maltreatment of dogs gradually came to an end, perhaps due less to pity than to another reason: in 1870 Lago di Agnano was drained to create more farmland and to combat the malaria-carrying mosquitoes. Without the water, it was no longer possible to revive half-suffocated, seemingly dead victims of the cave. Now, the green bed of the lake boasts a horse racecourse, the highway to Naples runs straight across it, there are high-rise tower blocks on the rim of the crater and on the south side, where the gas bubbles used to rise up through the water, there is a gigantic thermal complex, the Terme di Agnano. There the Neapolitans wallow in mud baths at the expense of their medical insurance. Around the thermal baths, there are 78 springs where boiling water bubbles up fiercely with the energy of a dog trying to save itself from suffocation. In the streams, lined with palm trees, precipitated bright orange iron crusts, gypsum crystals and alum flakes create a colourful spectacle. It must have been an astounding sight when the lake first dried up.

But now I want to see the cave itself. That proves not to be as simple as it sounds. It does not appear on my maps. The tourist office in Naples knows nothing about it; nor does the reception at the Terme di Agnano. It is not on their grounds. They point vaguely in the direction

The Grotta del Cane in 2009: the upper, dry part is filled with cold air, and the lower, wet section with warm CO_2. A man walking upright will feel no ill effects until he has reached the deepest part of the cave. A dog will not rise above the CO_2 level and will suffocate.

of the southeast side of the crater. I drive over there and find a small parking place alongside the road, closed off with a chain. Behind it is a large gate. Another gate – they're so afraid that someone will get into the Underworld. I can't see the cave through the gate. So there's nothing for it but to go back.

But I refuse to give up. On the Internet, I find the Associazione Conca di Agnano and its chairwoman, a doctor called Silvana Russo.

She is a very charming lady who immediately offers to accompany me to the cave, and we arrange to meet there. First, however, I meet the secretary Liborio Fusco at the parking place. He opens the chain and then the gate, by which time *la signora* Russo has also arrived, together with a young councillor from Agnano who has never seen the riches in his own locality. Together we walk along the path to the cave.

I would easily have walked past it. The cave is completely hidden from view by a thick tapestry of ivy, brambles and acacia branches that have grown down from the rock face and hang across the entrance. Apparently no one has been here for many years. Liborio has come prepared. He has a hockey-stick-shaped machete with him, which he uses to hack away at the vegetation. Gradually another gate appears. But everything's fine; the keys turn in the lock and it swings open.

It is smaller than I thought. It is an artificial tunnel, but I can hardly stand upright, though that is necessary to avoid suffocation. Perhaps it would be higher if the thick layer of earth and dead leaves at the entrance were cleared away. Silvana Russo tells me that it was cleaned up once, in 2001. It took two months to remove the tons of waste from the entrance. There were old shoes, bike saddles, bottles, cans, old clothes and masses of propaganda material for the 1968 elections. It all caught fire, and the heat was so immense that at first the fire brigade refused to go into the cave. The gate and the path were added after that, but everything has now become overgrown again. Tourists don't come here, nor do the guests at the baths.

The dividing line between the air and the carbon dioxide is very clear to see: the lower part of the wall is damp, warm and dark, while the upper part is dry, cool and light. The line is horizontal, while the passage slopes downward. The wedge of oxygen is therefore small, and if you walk further into the cave, you will be completely surrounded by the heavy CO_2 gas. That is not to be recommended: chemical analysis has shown it to be 70 per cent carbon dioxide and 30 per cent nitrogen, at a temperature of 78°C. I don't have a dog, but I did bring a lighter with me from the place I'm staying. I move it up and down, just as Caramuel and his fellow researchers did with their torch. And it's true: the flame dies out as soon as it goes below the dividing line. You don't even need a dog – unless of course you think it's arsenic. If Kircher had thought it through he would never have written that, because arsenic would not have extinguished the flame.

Rosario Varriale, the speleologist who led the clean-up operation in 2001, wrote in an article that he sent to me that he had penetrated

32 metres into the passage, wearing an oxygen mask. At the end, he found a chamber with *cocciopesto* on the walls, a kind of lime mortar typical of the Greek era, the second or third century BCE. This implies that the cave probably did not fill up with carbon dioxide until later. Varriale's camera exploded in the heat. Clearly no photos may be taken in hell.

What was the purpose of the tunnel? Perhaps it was a sudatorium, like those at the Solfatara? Or a failed attempt to excavate new thermal springs? Perhaps, for the men of the time, it was enough for them to stand in the gas up to their waists, Silvana suggests with a wink.

Now there are three craters – Averno, Solfatara and Agnano – and everywhere holes, emanations, hot springs. What is going on here? Where is the real entrance to hell? Let's take a look from slightly higher up. Then we see that there are many more: the Astroni, for example, the only crater still covered in dark forest and with a reed-lined lake at its centre, where the grebes paddle merrily back and forth. This was once the hunting ground of the Bourbons, but is now a nature reserve run by the Worldwide Fund for Nature, a lush oasis in this heavily built-up area. If you want to know what Aeneas saw in the forest of Averno before he entered the Underworld, take a look here. Then there is the Gauro, the Schiana, the Toiano, the San Vito,

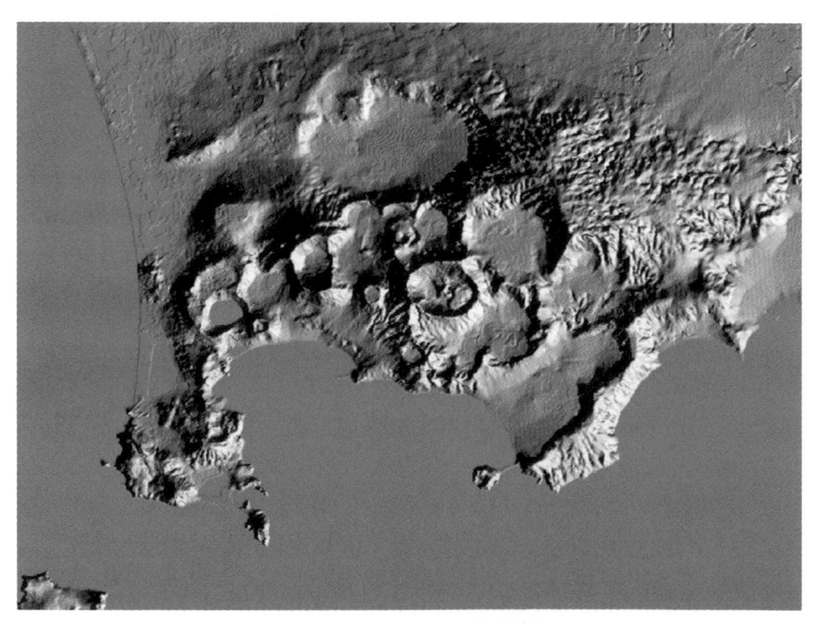

Radar image of the craters of Campi Flegrei made by the ERS-SAR satellite, 1993–6.

the Soccavo, the Quarto, the Monte Ruscello, the Cigliano, the Senga: every single one a crater! The detailed geological map by Rosi and Sbrana shows more than twenty. The whole area is so full of craters that an inhabitant of the moon would feel more at home here than anywhere else on Earth. So what is going on here?

That it is something to do with volcanic activity is clear: the area has been known as Campi Flegrei, the Phlegraean (Burning) Fields since antiquity. But the Swiss geologist Alfred Rittmann (1893–1980) was the first to understand the link between all the craters. According to his biographers Ippolito and Marinelli, he was something of a rebel. In the 1930s he was one of the few who steadfastly believed in Wegener's theory of continental shift, while the scientific consensus of the time dictated that continents could not move. Because of his rebelliousness, there was no place for him in the conformist scientific world of the time and when a rich Swiss banker asked him to set up a volcanological institute in Naples, he grasped the opportunity. His *Vulkane und ihre Tätigkeit*, published in 1936, was the first modern book on volcanology. It was reprinted countless times and was still a classic when I studied in the 1960s.

In an article in 1950, Rittman noted that all the craters of Campi Flegrei lie within a circle with a radius of around 12 kilometres. He concluded that the circle itself was a crater rim incomparably larger than all the small ones we have seen until now. It was the rim of a caldera, which comes from the Spanish word for cooking pot. Calderas, craters at least 10 kilometres in diameter, are not unusual. Those of Krakatoa and in Yellowstone Park are well known, while Italy also has a number of spectacular examples, including Lago di Bolsena, Lago di Vico and Lago di Bracciano. The caldera at Campi Flegrei was not discovered earlier because it lies on the coast. The beautiful semicircular Bay of Pozzuoli roughly follows the contours of the caldera, but there are so many semicircular bays in the world that it was not immediately apparent that this one happened to be a caldera. Furthermore, the southern part of the crater rim lies under water, well out into the bay.

Calderas arise as the result of exceptionally violent super-eruptions, during which dozens to hundreds of cubic kilometres of magma are ejected into the air. They often develop beneath an existing stratovolcano, comparable to Vesuvius, but the gas pressure under the volcano becomes so high that the almost the entire magma chamber is emptied in one massive eruption. When the energy of the eruption abates and the eruption column collapses, it creates pyroclastic flows that race

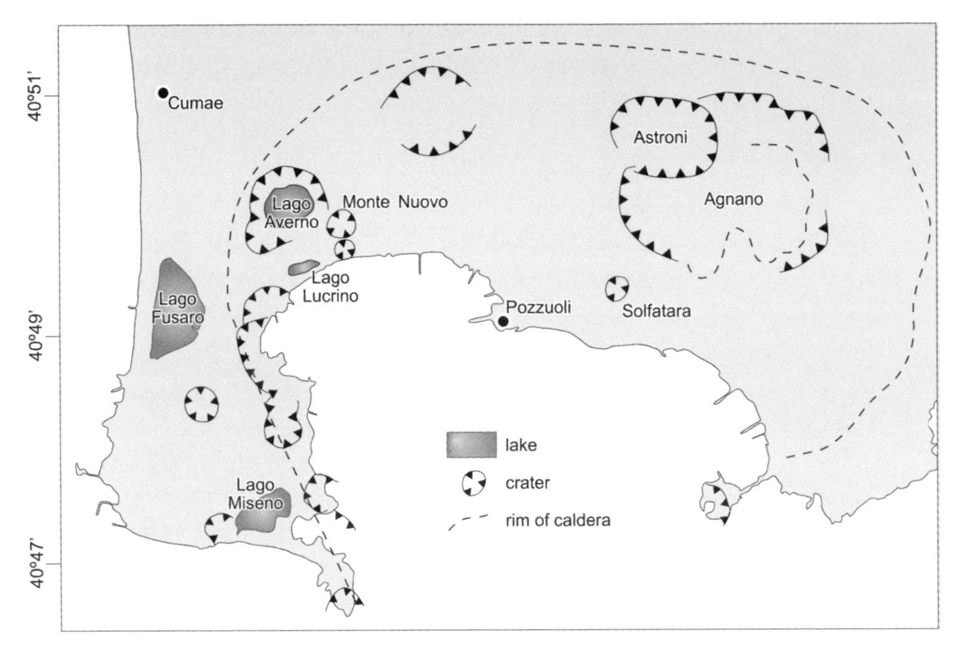

The caldera and the smaller craters of Campi Flegrei.

over the land at enormous speed and cover the entire landscape with thick layers of ash and pumice stone, known as ignimbrites ('fiery rock dust clouds'). The old body of the volcano collapses in its entirety into the empty magma chamber, leaving a large crater: the caldera.

Rittmann hit upon the idea that Campi Flegrei must also be a caldera because of the presence of ignimbrites. He distinguished two types: the Campanian ignimbrite or *tufi grigi*, grey tuff, and the *tufi gialli*, the yellow Neapolitan tuff. Each of the two layers of tuff represents its own eruption. The first created the large, 12-kilometre-wide hole, and the second left a new hole around 8 kilometres wide in the same place.

During the first eruption an area of 230 square kilometres collapsed. That comprises not only the current Campi Flegrei and the Bay of Pozzuoli, but also the whole city of Naples and the northwestern part of the Bay of Naples. An area of 30,000 square kilometres was covered with a thick layer of ash. The total estimated volume of the eruption was 100 cubic kilometres. Ash particles from the eruption have been found in the Mediterranean Sea as far away as the Nile Delta, in Central Russia, Turkey, and in archaeological excavations in Geissenklösterle in southern Germany.

It happened in the middle of the last ice age, some 38,000 years ago. As the ash cloud was drifting over Geissenklösterle, in a nearby cave

Homo sapiens was fashioning his first work of art from a mammoth tusk, the Venus of Hohle Fels. The recently discovered statuette, with its large breasts and prominent vulva, is considered pornographic by some.

Much later, only 15,000 years ago, when the last ice age was almost over, there was an almost equally violent eruption, of Neapolitan yellow tuff. The magma chamber under the caldera at Pozzuoli emptied itself again, covering an area of 1,000 square kilometres with a layer of ash up to 150 metres thick. Some 50 cubic kilometres of ash were ejected, half that of the Campanian grey tuff, but still an awe-inspiring amount. And because it lies on top of the Campanian layer, it is much more accessible. It is the foundation on which Naples is built, and of the capes of the peninsula of Posillipo and the city of Pozzuoli. We have come across it before: the tunnel at Lago Averno, the Antro della Sibilla, the Roman cisterns at Cumae and the Grotta di Cocceio were all hacked out of it. The Underworld itself provided Sybil with the material to become an underground oracle. The yellow stones used to build the *opus reticulatum* in her Temple of Apollo, on top of the plateau at Cumae, are made of tuff from hell. And if the whole area had not consisted of yellow tuff, Agrippa would not have been able to build such a large military complex.

Was that the end of it? No, not by a long way. At least another 70 eruptions followed. All the smaller craters that we can now see on the surface were created after the Neapolitan yellow tuff caldera. They were by no means such large eruptions, but we can still find layers of ash from each crater in the local area, and sometimes further away. Agnano, Solfatara, Averno, Astroni – none of them is older than 3,700 to 4,800 years old. Given that the story of Aeneas takes place after the Trojan War, probably around the twelfth century BCE (3,200 years ago), the 3,700-year-old Averno crater had hardly come to rest. The same was probably true in Virgil's time. That shows how closely human history pursues geological history around here.

And the story continues. Under the whole region, there is a large magma chamber at a depth of 4 to 6 kilometres, just as there is under Vesuvius. It is this magma that produces the sulfur and carbon dioxide, and warms the water for the thermal baths. What is more, under the volcano is a thick layer of limestone, the same rock of which the peninsula of Sorrento and the island of Capri consist. The magma reacts with the limestone, releasing carbon dioxide. And the more sulfur there is in the magma, the more explosive the following eruption will be. Mofettes are not indications of dying volcanic activity, as Berend Escher wrote: the volcano is at best holding its breath until it has gathered sufficient energy for the following eruption.

From 1982 Pozzuoli was startled by a large number (around 10,000!) of small-scale earthquakes, the heaviest of which, on 4 October 1983, measured 4.2 on the Richter scale. At the same time, the whole Campi Flegrei began to rise up, to almost two metres in 1984. There was widespread panic, and it was decided to evacuate 40,000 people from Pozzuoli to Monte Ruscello, a small volcano within Campi Flegrei. But there was no eruption and after 1984, the ground began to subside again. It was a bradyseism, a slow uplift and descent of the ground, probably caused by the expansion of overheated groundwater above a magma chamber. Since then, with short intervals, the ground has been subsiding again.

The periodic rise and fall of Campi Flegrei has been known for many centuries. The Romans had to regularly raise the level of the coast road to Pozzuoli because the ground was continually subsiding. And the 'Temple of Serapis' on the central square in Pozzuoli – which again is not a temple but a Roman marketplace (macellum) – repeatedly disappeared under water because of ground subsidence and had to be raised.

On the square, there are three columns of banded marble, the remains of a building destroyed by an earthquake. The columns record the history of 2,000 years of bradyseismic activity. At about two metres above the feet of the columns there are countless holes made by a kind of mussel, *Barnea candida*, that bores into clay, wood and soft stone. Some of the holes still contain the mussels. This is very strange, considering the mussels live only in seawater.

Italian mineralogist Scipione Breislak was the first to refer to this phenomenon, in his *Essais minéralogiques sur la Solfatare de Pouzzole* (1792), and wonders how it came about. Were the holes already in the stone when it was quarried? That is unlikely, since they are different pieces of rock from different quarries. Were they inundated by the sea after an earthquake and then raised above the water level again by a second quake? But then, together with the walls of the temple, they would have been knocked down by the vibrations. He eventually concludes that they probably came from an older temple that now lies under water.

It was Charles Lyell, the pioneer of modern geology, who came to the right conclusion when he visited the site in 1828: the land first subsided, then rose again, and not as the result of rapid earthquakes, but of slower processes. He even used an engraving of the macellum

The three columns on the flooded macellum in Pozzuoli (the 'Temple of Serapis'). The slightly rougher part of the columns, affected by the mussels, is about a third of the way up.

on the front page of his book *Principles of Geology*, published in 1830. He could not have found a better illustration of his theory on the length of geological time. With broad gestures, he extrapolates this to other parts of the Earth, but we now know that it is only a local phenomenon: bradyseism of the Campi Flegrei caldera. The Underworld is alive, and now and again it pushes Pozzuoli up into the air, much to the alarm of its residents.

Actually, we've already encountered these effects elsewhere: they explain why the Portus Julius of Octavian and Agrippa now lies 12 metres under water, why the canal between Lago Averno and Lago Lucrino fell into disuse, and why the tunnel with the red gate, the fake Grotta della Sibilla on Lago Averno, is flooded. It may not have been the Styx, but it was the fault of the Underworld.

It did not always go as slowly as Lyell thought. After Roman times, the ground continued to subside steadily at a rate of little more than a centimetre a year, but after 1503 Pozzuoli suddenly started to rise: seven metres in 35 years! The columns in the macellum rose majestically higher and higher from the sea until, in 1538, there *was* an eruption. Within a week, it had created a new volcanic cone 100 metres high, Monte Nuovo, right next to Lago Averno.

The story is beautifully told in the standard work by Sir William Hamilton dating from 1776, *Campi Phlegraei: Observations on the Volcanos of the Two Sicilies*. The book was written 240 years after the eruption, but is based on eyewitness accounts by Marco Antonio delli Falconi (1538) and Pietro Giacomo da Toledo (1539). William Hamilton was a diplomat, His Britannic Majesty's Envoy Extraordinary and Plenipotentiary at the Court of Naples, with a keen interest in volcanoes. His book comprises the letters he sent back to the Royal Society during his stay in Naples, and includes the most splendid colour engravings ever produced on the city and its environs, drawn by Pietro Fabris. They show the eruption of Vesuvius in 1767, with the lava flows and ash clouds, and a timetable of the incidents in diagram form. There are also illustrations of older lava flows, layers of ash from earlier eruptions, very detailed drawings of the various volcanic rocks, the Solfatara, Astroni, Lago di Agnano before it was drained (but without the dog), our own Lago Averno – and the crater of Monte Nuovo. He wrote:

> The eruption made its appearance the 29th of September 1538, the feast of St Michael the angel; it was on a Sunday, about an hour in the night; and, as I have been informed, they

> began to see on that spot, between the hot baths or sweating rooms, and Trepergule, flames of fire, which first made their appearance at the baths, then extended towards Trepergule, and fixing in the little valley that lies between the Monte Barbaro and the hillock called del Pericolo (which was the road to the lake of Avernus and the baths), in a short time the fire increased to such a degree, that it burst open the earth in this place, and threw up so great a quantity of ashes and pumice stones mixed with water, as covered the whole country; and in Naples a shower of these ashes and water fell a great part of the night. The next morning, which was Monday, and the last of the month, the poor inhabitants of Pozzuolo, struck with so horrible a sight, quitted their habitations, covered with that muddy and black shower, which continued in that country the whole day, flying death, but with faces painted with its colours; some with their children in their arms, some with sacks full of their goods; others leading an ass, loaded with their frightened family, towards Naples; others carrying quantities of *birds of various sorts, that had fallen dead at the time the eruption began* [my italics]; others again with fish which they had found, and were to be met with in plenty upon the shore, the sea having been at that time considerably dried up.

In the days that followed there was a series of further explosions, but a week later it was all over. The Italian volcanologists Mauro Di Vito and Claudia D'Oriano and their co-authors carefully reconstructed the events surrounding the eruption of Monte Nuovo, the only one in Campi Flegrei in recorded history. Their reconstruction shows that before the eruption, Lago Averno was again linked to the sea, as a result of the same ground subsidence that the mussels had taken advantage of in Pozzuoli. Only after the eruption was the lake once more cut off from the sea, as a result of the rapid uplift in the land and the ash flows from Monte Nuovo, and it has remained so ever since.

The eruption started near the thermal baths at Trepergole. 'This water is very useful and cures the sweat caused by a weak stomach', states the Neapolitan *Trattato dei Bagni* from around 1220. But the stomach of the Earth ejected something a lot stronger than water alone. Was this Aeneas' 'foul jaws of stinking Avernus'? Did Virgil witness a minor eruption at the small mountain known as Il Pericolo? Is the entrance to the Underworld now buried under the ash of Monte Nuovo?

Copia de Una lettera

di Napoli che contiene li stupendi, & gran prodigij apparsi sopra à Pozzolo

The eruption of Monte Nuovo over the Bay of Pozzuoli, 1538, engraving.

I have to climb Monte Nuovo, that much is clear. The footpath to the top starts at Lago Averno. I drive along the same road as before, but notice only now that the tunnel I have to pass through to reach the access road cuts straight through Monte Nuovo. The Romans bored through the old tuff, the Italians through the new. As I drive through the tunnel, I look to the left and right in the hope that I might see traces of the buried jaws of stinking Avernus, the small Il Pericolo or the baths of Trepergole. But everything has been hidden behind the tunnel wall. A little respect for the history of the underground is asking too much.

I park the car again at Bar Caronte and walk up through the plum orchards. There is nothing to suggest that this mountain is less than 500 years old. Halfway up the hillside, I turn around to find I can look out across the whole of Lago Averno. It is beautiful, charming and heavenly. The lake, too, is only 4,000 years old. But what has it gone through since the time of Virgil? Deforested, cultivated, two tunnels and a canal through the crater wall, twice connected to the sea and twice cut off again, subsided, lifted 7 metres into the air and dropped again by bradyseismic activity, half filled with ash from the eruption of Monte Nuovo, which perhaps also put an end to its thermal baths, though Rittmann did observe bubbles rising the surface of the lake in 1950; no wonder that the entrance to hell is no longer to be seen.

On top of Monte Nuovo, I look down into a magnificent, deep, young crater. According to Rosi and Sbrana's map, it still emits hot gases. It would have made a perfect entrance to the Underworld, except that it wasn't there in Virgil's time. I don't descend into the crater, but walk around it. On the southern side of the heavily vegetated rim, a fantastic view of the sea suddenly opens up: the Bay of Pozzuoli.

I suddenly realize that only now am I looking at the real crater: the enormous Campi Flegrei caldera, 12 kilometres across, with its centre in the bay. It makes no difference where Aeneas entered the underground – at Lago Averno, Solfatara, the Dog Cave, Astroni crater or Il Pericolo – because under the ground they all come together in the same magma chamber. They are all small, bubbling outlets of the same immense volcano. Hell is a little larger than Virgil imagined.

FIVE

The Vestibule

Thick hail and dirty water mixed with snow
come down in torrents through the murky air,
and the earth is stinking from this soaking rain.
Dante Alighieri, *Inferno*, VI:10-12

My mother loved the forest. My father always wanted to go the mountains, but when I was a child, my mother always won and we would go to a narrow strip of woods on the Brabantse Wal, on the border between the provinces of Brabant and Zeeland. On my father's old linen ordnance map, it was called 'Platluis'. At first we just used to go there for a picnic on Sundays. My dad used to remove the backseat from the 2CV, and we would sit on it on the soft blanket of pine needles and eat our sandwiches. Later, she bought a plot of land there herself, with fir trees bent over by the wind, and built a small wooden cabin on it. It was here that I started my first fieldwork into the Underworld. I would dig a shallow pit in the sandy ground, half cover it with planks and sand, and lie in it so that no one could see me from the house. I was only ten years old, and the associations with the grave and with death did not occur to me. It was a multicoloured pit: the top layer of earth was black, the one under that lead-grey, and under that it was rusty brown and hard, with fine, capricious bands running through it. I know now that it is called podzol, from the Russian *pod-zol*, meaning 'under the ash': not because of the ash of cremated bodies, or from the ash drawer of Russian tile stoves, but because of its ash-grey colour. At the time, I didn't pay it any special attention.

Some 25 years later, I found myself among people who had devoted their entire lives to it. At Wageningen Agricultural University there were professors of soil genesis, soil fertility, soil physics, soil suitability, soil evaluation and so on. There was even a geologist whose job was to unravel the significance of the underground for the soil. That was me.

Wekeromse Zand, Veluwe, the Netherlands. Podzol covered with a thin layer of light drift sand; the extreme outer layers of the gobstopper.

The study of soil, pedology, starts with a pit. Not shallow like the one I used to hide in near our cabin, but about 2.5 metres deep. The idea is to obtain a clear vertical cross-section of the ground, a soil profile, showing exactly what layers, 'horizons', make up the soil. Digging such a pit is an art in itself. It has to meet a number of criteria. First of all, you have to be able to get in and out of it. At 2.5 metres deep, that means you will need to make steps in it. In addition, it has to be wide enough for you to face the wall and easily make observations and take samples, and to bend over to look at the lower layers. The wall has to be made smooth enough with a spade to see small variations in colour without disturbing the natural structure of the soil. Pedologist Jan Bokhorst recounts his memories of field trips he made as a student in the story '*Graaf vaker een kuil*' (Dig a Pit More Often): 'The enormous profile pits the Poles had dug for us were impressive. They could accommodate several people simultaneously and you were immediately confronted with the soil and the opportunities and limitations of earthworms and the roots of plants to develop.'

In pits like these, generations of pedologists armed with black-and-white silhouette cards have determined the form, size and structure of clumps of soil and filled in their measurements on large forms. A tape measure hangs down the side of the pit, with the zero at ground level, preferably with the decimetres marked in red, green and yellow, so that on the compulsory photo you can clearly make out the depth of the soil horizons. Pedologists always have a trowel or a knife to hand so that they can dig around in the ground, a magnifying glass to observe the fine structure, a flask of hydrochloric acid to determine the carbonate content of the soil, and a book of colour samples to determine the colour of the soil according to the Munsell Scale. Under each colour there is a small hole in the hard, grey card. You can wheedle a small lump of soil into the hole to compare the colour with that on the card. Perhaps the most important measurement is the texture, the grain size composition: how much sand, silt and clay the soil contains. Experienced soil surveyors roll a small clump of soil into a sausage and stick it in to their mouths, chew it, ponder it for a while, and then pronounce their verdict: light loam, 15 per cent clay.

And then it starts, as there is always more than one pedologist around the pit at any one time. They take turns to jump in, pick at the wall and taste the soil: 'I think it's heavy loam, 20 per cent clay.' They will disagree about the depth of the humus layer, and how the soil originated: has iron leached down from the lead-grey layer into the brown layer beneath it, or is it just humus, or have thin streaks of clay leached into it, and how and when did that happen? There are endless discussions on the basis of qualitative, subjective observations,

'Bullshit around the pit', Turali, Dagestan.

where the lack of statistical evidence is compensated for by years of experience in hundreds of pits. Bullshit around the pit, that's what we call it.

Then it's time to take samples. Brown, double-walled paper bags are filled with soil, at least one for each layer, and it's up to the soil laboratory to decide who was right. Sometimes you push aluminium tins into the wall to make a microscopic section and, if it's really worth it, you impregnate the whole wall with polyester lacquer, stick a strip of medical gauze to it, pull it off again with the layers of soil stuck to it and put it in a wooden frame to take to the laboratory. The soil museum at Wageningen is full of these soil monoliths.

Since you can't really ask a farmer to let you dig pits all over his fields, if you want to study variations over a wide area, you have to adopt a less rigorous approach. What you need then is an Edelman auger, a hand tool with a drill head consisting of two parallel sharp blades running to a point at one end, and two handles at the other, which you use to twist it into the ground. The first Edelman auger from 1943 was 1.2 metres long; according to pedologists, this was the depth of the soil, that is, that layer of the Earth's crust that can be reached by the roots of plants. Wageningen professor Pieter Buringh, who studied under Edelman, claims, however, that it was this long so that it fitted exactly in the back of a Volkswagen Beetle.

The soil is actually a very special part of the Earth. It is the outermost skin of the gobstopper, the place where earth, air, water and life are so intimately entwined that it has become something autonomous. It is the best example of a whole that is more than the sum of its parts; an ecosystem in itself, concealing more hidden life than you would ever suspect. And that life is essential: my former colleague Nico van Breemen once created two soil profiles between two vertical perspex sheets and displayed them in the hall of our institute. Both consisted of a layer of clay with a layer of leaves on top. He introduced worms to one of the profiles and not to the other. The difference was astounding. Within a couple of months the worms had mixed the leaves and the soil together, creating real soil, while nothing visibly happened in the other at all.

Some people claim that there is more biodiversity in a clump of garden soil than in the entire Amazon rainforest, while there is almost three times as much carbon dioxide in the ground as in the whole atmosphere. What is more, it can sometimes remain sealed in the soil for thousands of years in stable humus particles.

But the soil is also the part of the Earth's ecosystem that has been changed the most by human activity: we plough it, sow seeds in it, fertilize it, irrigate it, drain it, harvest it, graze it, divide it up into plots, level it, drive all over it in heavy agricultural machinery and pollute it with heavy metals. There is almost nowhere on Earth where the soil can still be observed in its natural state. That is the tragedy of pedology: it thrived due to agriculture, but at the same time agriculture has changed the soil irrecoverably.

Soil was only discovered to be a separate natural phenomenon in 1883. None of the great scholars of classical antiquity, from the heyday of Arab learning or during the Enlightenment saw anything of interest in the subtle colour gradations just under the Earth's surface. Russian geologist Vasily Vasili'evich Dokuchaev laid the foundations for soil science in his thesis on the black soil of Russia. He understood that the black earth was so rich in humus because worms, moles and other creatures had mixed the digested roots of the steppe grass with the calcium-rich loess soil over centuries. Dokuchaev claimed that the black earth was more valuable than the mineral resources in Siberia, because it was the most fertile soil in the world. But the cultivation of the steppes for agriculture at the end of the nineteenth century exposed much of the humus to the air, causing it to oxidize. The fertility declined and agricultural production decreased. That was the reason for Dokuchaev's research.

It is an impressive achievement to elevate something that every farmer sees each day alongside the tracks of his cart, every gravedigger in his graveyard, to the status of a distinct natural phenomenon. Dokuchaev also discovered that the character of the soil depends heavily on the climate: if you travel from the cold north of Russia to the warm south, you see the soil change along with the vegetation. In 1889 Dokuchaev took a soil monolith of the black Russian earth and other soil types to the World Exhibition in Paris (the one for which the Eiffel Tower was built) and was awarded a gold medal. The monolith was partially destroyed during the student uprisings in the Sorbonne in 1968, but a piece of it can still be seen at the Institut National de la Recherche Agronomique in Paris.

Kalmykia, July 1991. The sun is beating down mercilessly on my tent. The canvas is covered with prints of headlines from *Pravda*, *Izvestia* and *Trud*, not to mention *L'Unità*, *Libération*, the *Morning Star* and even the Dutch daily *De Waarheid*, so I am greeted every morning by Communist press from around the world. I look outside.

Soil monoliths in the Museum of Natural History, Moscow State University.

The immense, undulating steppes, actually semi-desert, exude the penetrating aroma of wormwood in bloom. In the middle of this wilderness stands the *wagonchik*, the square cart belonging to Yevgeny Vasil'evich Tsutskin, the bearded archaeologist who is going to tell us about his excavations. The physical presence of Karl Ernst von Baer can be felt here: the elongated Baer hills survey Tsutskin's pits from the horizon with a heavy frown. For breakfast, Tsutskin has made us a big pan of *ukha*, fish soup, in the small kitchen on his caravan, a shoebox on wheels that looks more like an enlarged toy than the luxurious cocoons that are hitched to Western Europeans' cars when they speed off down the motorway for a holiday in the sun.

With 27 students from Wageningen and eleven Russian counterparts, we are following in the footsteps of Vasily Dokuchaev from Moscow, via the Caspian Sea, to the Caucasus. We have seen the loess soil change through the different climate zones from podzol-like earth in the pine forests to the black earth of Dokuchaev's long-grass steppes. We have seen the long windbreaks of trees planted in the steppes on his recommendation after the great drought of 1891 to stop the agricultural soil from drifting. But he felt that part of the steppes should be preserved in its original state. This led to the foundation of the first Russian nature reserve in 1892: Kamenny Step, meaning 'Stone Steppe'. Tolstoy and Chekhov were among those who collected money for the project. A century later, the black earth is still almost virgin.

Now we are almost a climate zone further: here in the semi-desert of Kalmykia, rainwater no longer washes chemicals downwards into the soil, as it does with podzol. Evaporation causes a reverse process: the groundwater seeps up through capillaries, depositing calcium, gypsum and even salt in the soil. This is nomad country: the Kalmyks are Buddhist nomads originating from Tibet. In the seventeenth century, during the last Mongol invasion of the West, the Tsar gave them permission to settle here. Stalin deported them in the Second World War, but they were later allowed to return. However, Tsutskin is not interested in the Kalmyks but in the people who lived here 4,000 years ago and who left *kurgans* (burial mounds) everywhere in the desert. In his pit in one of these mounds, he shows us how they were constructed on top of the virgin ground. They did not want to dig into the soil, he explains, because they considered Mother Earth too valuable. 'Do you know why Turkish slippers have those curled-up toes?' he asks, and eagerly answers his own question: 'For the same reason: so that the points of the shoes cannot damage the earth.' I thought this such a wonderful idea that the image has always remained with me.

We have been unable to hold on to that principle. By using the soil, humankind has saddled itself with impossible dilemmas. During the cultivation of marshy deltas, so much sulfuric acid formed in the soil that the crops could not cope with it; this has become an insoluble

Turkish slippers.

problem for many rice farmers in the tropics. Farmers call this soil 'cat's clay' because it is full of yellow stains the colour of cat's urine.

During the building of a new residential estate in Baarlo, the Netherlands, in 1988 the soil was discovered to contain extremely high levels of arsenic, eight times higher than the legal limit. Homes were evacuated and the toxic ground removed. Had some company or other dumped poisonous chemicals here in Baarlo illegally? Not at all: the arsenic concentrations were natural, dissolved from pyrite, iron sulfide, in old sediments deposited in the soil, along with iron ore, by groundwater flows. Was nature itself breaking the law? Tests with 'simulated gastric acid' (diluted hydrochloric acid) showed that the arsenic posed no danger to public health. So the building went ahead. Otherwise some 50,000 hectares of similar soil in other parts of the country would also have had to be removed.

The soil around the abandoned zinc mine at Moresnet, near the point where the Netherlands, Belgium and Germany come together, has unacceptably high levels of zinc. The soil should actually be decontaminated here, too, but there is a rare variant of the mountain pansy, the subspecies *Calaminaria*, which thrives on the high zinc content. If the soil is cleansed, the pansy will disappear.

The most bizarre example of soil contamination must come from China. Extremely high levels of mercury have been found in a hill near the city of Xi'an. The hill, which covers two square kilometres, is not natural; nor is it a slag heap from a mercury mine, the remains of primitive gold-mining activities or the final resting place of used mercury thermometers. It is in fact the final resting place of the legendary emperor Qin from the third century BCE – at least, that is the theory. The hill has not yet been excavated; the mercury-rich soil samples come from bores.

One of the world's greatest archaeological treasures was found on the edge of this hill: the famous terracotta army. In his *Shiji*, the Records of the Grand Historian, the Chinese scholar Sima Qian (145–90 BCE) tells us that the emperor Qin had a miniature replica of his entire kingdom constructed in his copper tomb, with model palaces, pavilions, offices, a domed ceiling depicting the firmament with precious jewels depicting the stars, Sun and Moon, and of course a model Underworld. The great rivers in his kingdom, the Huang He and the Yangtze, were replicated too, running with mercury down to a miniature mercury ocean, the whole thing regulated by some kind of mechanism. There are crossbows standing at the ready, designed to go off automatically if grave robbers should try to break into the tomb.

The *Shiji* contains the only description of what was reputed to be under the ground, and the extremely high mercury content in the soil suggests that it may be accurate. As yet, however, no archaeologist has dared to dig any further, not because of the mercury, but out of fear of causing irreparable damage.

Today's soil science focuses on cleaning up contaminated ground. Classical pedology around the pit has practically disappeared, the whole world has been pedologically charted and enormous agricultural databases advise farmers where and when they should plant their crops. How many farmers still know what cat's clay, quick clay, Rodoorn soil or gliede (greasy humus) are? Farmers adapt the fertilizer dosage to the soil type, which they can determine using GPS and digital soil charts. Thousands of municipal officials draw up zoning plans and environmental impact reports for land use on the basis of current data, but out in the field they can hardly tell podzol from marine clay. Pedologists from around the world complain in a booklet devoted to the subject, *The Future of Soil Science*, that there is no longer any money for their research. Apparently we already know all there is to know.

At Wageningen, the memories of the podzol from my youth came back. Another pit, more podzol, but this time a coffin was lowered into it, containing the body of Jacob Bennema, my colleague at the university, professor of soil classification and land evaluation. He had a PhD in rising sea levels, but in his final years he had devoted his time to soil suitability. I considered it very appropriate for a professor of soil suitability to be buried in such beautiful soil. Some people want a splendid coffin, their hair neatly combed and to be dressed up in their best suit or dress, with nice speeches and music, but I want to be buried in good soil. I have stood by graves in marine clay, rather dull and sombre, which sounds so heavy as it falls on the coffin. Podzol is much better: the sound off the sand is lighter, which is why pedologists talk of heavy clay and light sandy soil. There is more air in sandy soil, so that you decompose and become one with the earth more quickly. In their advertisements, funeral directors do not devote sufficient attention to the quality of the soils they can offer.

The soil, that is the hell we really end up in. In our journey to the centre of the earth, we come no further than about 1.5 to 2 metres below the surface. If you are afraid of that, you can always be cremated. Then you immediately become CO_2 and disappear into the atmosphere. You go straight to heaven, but you are lost to the underground. All you can do is hope that there is a tree somewhere that is willing to absorb your atoms somewhere in the process of photosynthesis, otherwise you might end up in an air bubble in the Antarctic ice cap and then you'll be out of circulation for the next million years or so. There is a small pile of ash left over, but not much. Some people have their ashes compressed into a diamond. That might ensure that you are preserved for eternity, but also has its risks. Imagine your daughter accidentally dropping you down the plughole in the sink.

If you are buried it can take hundreds to thousands of years before you become CO_2. You get the same treatment as the roots of the steppe grass, mixing slowly with the soil until you once again become one with the earth – so long as your grave is not prematurely cleared.

There will at least be someone who will feel sad if your grave is cleared: the archaeologist of the future. Within a few decades, the dearly departed become a valuable source of archaeological information. Archaeologists are soil scientists of a sort. They also dig pits, but they are different: they are especially interested in the horizontal dimension as well as the vertical structure of the soil. Pity the farmer on whose land an important settlement is found; archaeologists are not satisfied with a few hand bores. They will excavate hectares of his fields centimetre by centimetre to try and reconstruct how his predecessor lived in the Middle Ages or the Stone Age. With dozens of volunteers armed with trowels, artists' paint brushes and dustpans and brushes, they will dig painstakingly around in squares of ground delineated with kite string to measure each piece of pottery or flint, identify it and study it until they have discovered what it was originally used for.

While pedologists reconstruct the interplay of earth, water, air and life in soil profiles, archaeologists seek evidence of the human lives of the past. That is why they like to find graves. In most cultures life does not end when a body is buried in the soil. That is only the start of the deceased's journey to the Underworld – of which the soil itself is not considered a part. You don't physically go to hell, only your soul, your spirit and perhaps your carbon dioxide gas, though no one knew about that in former times. Yet you apparently need all kinds of earthly things to take with you on the journey. Otherwise what is the point of all those burial offerings – the beautiful robes, golden masks, amphoras and implements, the extravagant copper mausoleum, the mercury, the jewels and the terracotta horses and soldiers? What image did people have of the Underworld? Archaeologists seek answers to these questions. Excavations in the soil are a window into hell.

From dearly departed to interesting artefacts. Emergency excavations on the market square in Delft, the Netherlands, 2004.

The most symbolic burial offering is the obol. Obols were cheap coins that the Greeks and Romans, from the fifth century BCE to CE 1000, placed in the graves of the dead. The coin was placed in the mouth because that was the closest orifice to the soul. The soul had left the body and the obol prevented it from returning. Archaeologists find obols in skulls in ancient graves not only in Greece and Rome, but also as far afield as Spain and England.

In the Underworld, you have to cross the River Acheron, and you need money to pay Charon, the ferryman. Without an obol, you cannot cross, as the Greek-Syrian poet Lucian of Samosata explained

Obol, 3rd century CE, from Pantikapaion (Kerch) on the strait between the Black Sea and the Sea of Azov, Homer's entrance to the Underworld, according to Karl Ernst von Baer. It was customary to place a coin in the mouth of the dead in Kerch until the 19th century.

in his satirical *Peri Penthous* (Of Mourning), written in the second century CE. Acheron, Lucian said 'is too deep to be waded, too broad for the swimmer, and even defies the flight of birds deceased', a striking reminder of Lago Averno. Some of the dead even have two obols, perhaps in the hope of buying a return ticket.

The delightfully plebeian erotic novel *The Golden Asse* by Apuleius, dating from the same era, tells the story of Eros (Cupid), the son of Venus, falling in love with Psyche. Venus, jealous of Psyche's stunning beauty, kidnaps Cupid and will only allow Psyche to see him if she completes a number of impossible tasks. For example, she has to go the Underworld and ask Persephone for a box containing a piece of her beauty. Psyche is in despair and wants to cast herself from a tower, but the tower gives her good advice:

> By and by thou shalt come unto a river of hell, whereas Charon is ferriman, who will first have his fare paied him, before he will carry the soules over the river in his boat, whereby you may see that avarice raigneth amongst the dead, neither Charon nor Pluto will do any thing for nought: for if it be a poore man that would passe over and lacketh money, he shal be compelled to die in his journey before they will shew him any reliefe, wherefore deliver to carraine Charon one of the halfpence (which thou bearest for thy passage) and let him receive it out of thy mouth.

The tower tells Psyche that she must not look in the box on her way back. I'm sure you can guess the rest . . .

The tragedy is that the deceased who have been found in their graves with an obol in their mouths seem to have got stuck in the queue for the ferry, as they still have their tickets with them. Their journey to the Underworld was cancelled without explanation. Perhaps only those without obols in their mouths were carried across by Charon, leaving their physical remains behind to decompose in the ground. Perhaps they too followed Lucian's good advice: don't put an obol in your mouth, because then you may be able to return to the land of the living. Or perhaps they have become the eternal victims of grave robbers: no one knew better that burial offerings did not physically accompany their owners to the Underworld.

Charon's ferry service is not reliable; there are long lines of ghosts waiting to be carried to the other side. Some are not permitted to cross at all: according to Virgil, these are the dead without graves, like Palinurus, Aeneas' drowned helmsman. They remain at the gate to the Underworld until someone finds their bones and lays them to rest. Virgil says nothing about paying the ferryman, but it is logical that those who die by drowning have no grave, and therefore no obol.

Once you have made the crossing, you are in the Underworld but not yet in hell. After all, why give your departed loved ones money to pay for the ferry if it would only take them to hell? Those left behind knew that there was also the chance of a happy ending. The road only divides on the other side of the water: the bad go to the left, through the gate of Tartarus, but the good go to the right, to the Elysian Fields. In antiquity Paradise was also under the ground; it was just a matter of taking the right turning.

Aeneas was good and therefore, like Odysseus before him, never actually descended into Tartarus. Although, when they arrive at the iron gate, Sybil tells him about all the terrible ordeals that sinners undergo there, Aeneas himself sees none of it. As a result, we hear precious little from Virgil about its geography. Where the road splits, Aeneas goes to the right, the way of the gods and the wise, to the Elysian Fields, the Champs Élysées!

> They came to the pleasant places, the delightful grassy turf
> of the Fortunate Groves, and the homes of the blessed.
> Here freer air and radiant light clothe the plain,
> and these have their own sun, and their own stars.

> Some exercise their bodies in a grassy gymnasium,
> compete in sports and wrestle on the yellow sand . . .
> Look, he sees others on the grass to right and left, feasting,
> and singing a joyful paean in chorus, among the fragrant
> groves of laurel, out of which the Eridanus's broad river
> flows through the woodlands to the world above. (*Aeneid*,
> VI:638–43, 656–9)

Who would ever expect to find such a delightful place under the ground! You'd be happy to pay an obol to go there! Yellow sand! Fortunate groves! Feasting on the grass! It sounds a little like my own Platluis.

> And now Aeneas saw a secluded grove
> in a receding valley, with rustling woodland thickets,
> and the river of Lethe gliding past those peaceful places.
> (VI:704–6).

The Lethe is the river of forgetfulness. People have to first wash away their sins and forget, and then:

> All these others the god calls in a great crowd to the river
> Lethe,
> after they have turned the wheel for a thousand years,
> so that, truly forgetting, they can revisit the vault above,
> and begin with a desire to return to the flesh. (I:748–51)

It is fascinating that Virgil himself, after a bath in the Lethe lasting ten centuries, suddenly returns in an new role, as Alighieri's guide through the Inferno in his *Divine Comedy*. Virgil is Dante's role model as a poet, just as Homer was for Virgil.

But Dante has significantly changed the topography of the people waiting at the gate of hell to cross the Acheron in Charon's ferry.

> Here sighs and cries and shrieks of lamentation
> echoed throughout the starless air of Hell;
> at first these sounds resounding made me weep:
>
> tongues confused, a language strained in anguish
> with cadences of anger, shrill outcries
> and raucous groans that joined with sounds of hands,

raising a whirling storm that turns itself
forever through that air of endless black,
like grains of sand swirling when a whirlwind blows.
(III:22–30).

Strangely enough, Virgil does not seem surprised. This is no longer an assembly point for those who have drowned and others who have missed the boat. No one is here of their own free will; no one has been given an obol by their grieving family, unless they seriously wished to see the deceased spend eternity in hell. Those waiting here are all sinners, awaiting their punishment in hell, or doomed to remain on the jetty forever. The good have already been separated from the bad. The eager anticipation in the vestibule of Virgil's Underworld has completely disappeared, as has its own sun and stars. The air is now black and Dante rustles up a sandstorm to depict the tumult.

Why? What kind of people are now standing in line for the ferry? They are the cowards, the indifferent, those who could not choose

Gustave Doré's image of Charon herding the sinners on to his boat: who are they?

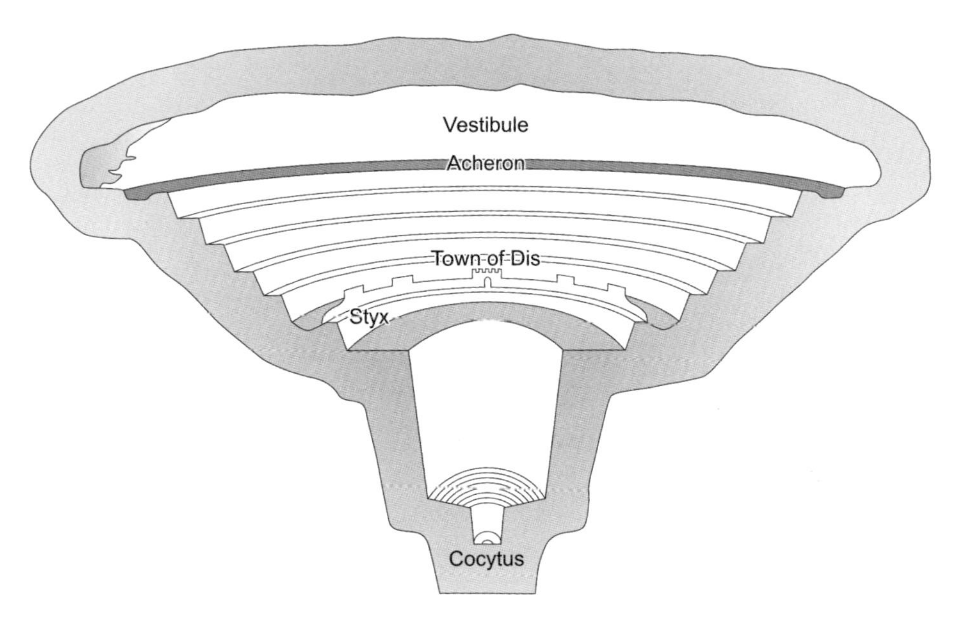

The vestibule of the cowards and the indifferent in Dante's *Inferno*.

between God and Satan. As some observers note, their punishments seem a little out of proportion to their crimes. At the beginning of his epos, Dante may have overstated in advance the horrors that he would have to describe later.

I have great sympathy for people who cannot choose. My father was a liberal, my mother a socialist. Both were party members and, during elections, posters for both hung in our windows. My father was of Jewish origins, my mother was Protestant. My father loved the mountains, my mother the forest. That taught me to keep all my options open. Bury me in that vestibule, the earth at Platluis. I already know the Underworld there, and I won't have to make a choice. But please put an obol in my mouth. After all, you never know.

SIX

Charon's Ferry

'All this will be made plain to you
as soon as we shall come to stop awhile
upon the sorrowful shore of Acheron.'
Dante Alighieri, *Inferno*, III:76-8

So you are sitting in the boat with Charon, you've paid your obol, but which river is this? According to the Greek poets it is the Acheron, but for Roman and modern-day poets it is the Styx. Why is that? Where are its source and mouth? And why does it flow underground? For someone like me, who lives in a delta, that is incomprehensible. We have groundwater, of course, but not in the form of rivers that you can sail boats on.

Let us start at the beginning. Homer, in the eighth century BCE, does not mention Charon. Nor does his Odysseus cross a river: instead he comes from the sea to a place 'where the rivers Pyriphlegethon and Cocytus (which is a branch of the River Styx) flow into Acheron, and you will see a rock near it, just where the two roaring rivers run into one another'. That the Cocytus is a tributary of both the Acheron and the Styx does not sound completely logical, but the Acheron is clearly the main river.

In the fifth century BCE the two authoritative Greek historians Herodotus and Thucydides, the first real historians of Western civilization, mentioned the Acheron; not in a mythological account but in a factual description of the region of Thesprotía, in Epirus on the west coast of the Greek mainland. There is no mention of Charon in these historical accounts, but there is an association with death: Herodotus refers to consulting the Necromanteion, the Oracle of Death, on the River Acheron in Thesprotía. That is still there, as we shall see. For the sake of convenience, Karl Ernst von Baer seems to have ignored the fact that a real Acheron exists. After all, no one can go everywhere.

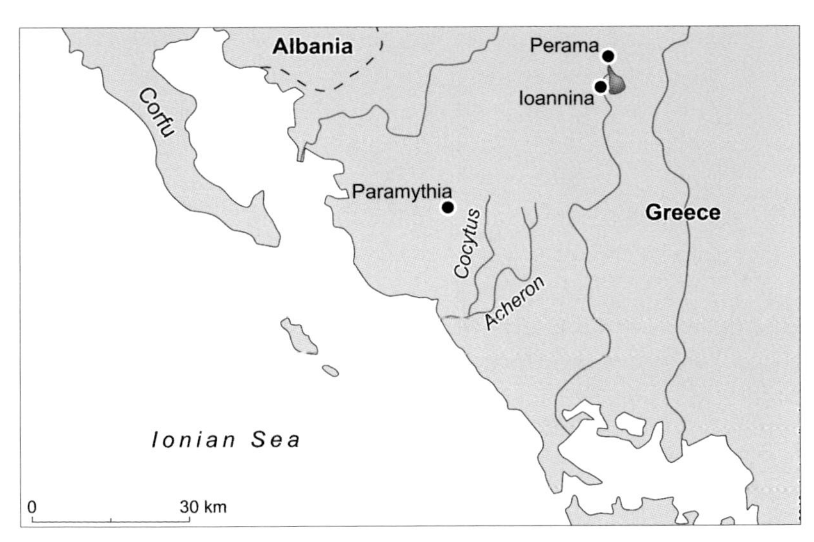

The Acheron and the Cocytus in Epirus, western Greece.

In Aristophanes' comedy *The Frogs*, from 405 BCE, Charon does appear as the ferryman of the Acheron, in a hilarious discussion with Heracles and Dionysus. Dionysus wants to go to the Underworld to bring back the playwright Euripides, who died in 406 BCE, and asks his brother Heracles the way, since he has already been there. Heracles tells him that the best way to get to the Underworld is to kill himself. But Dionysus says he wants to come back and follow the same route that Heracles took. His brother answers that it is a long journey that begins with a very large, bottomless lake. An old sailor takes you across in his boat, which is tiny, for the price of two obols. You will see serpents and a thousand monsters and, after that, a pool of eternally boiling mud.

Dionysus does not allow this to discourage him and sets off for the Underworld with his servant Xanthias. They come to the lake and greet Charon, the ferryman of the Acheron, who is mooring his little boat. 'Who seeks a rest from work and trouble? Who's heading for Fields of Forgetfulness, Never-never land, the Cerberians, the Ravens and Tartarus?' Charon asks them. 'That's me', Dionysus replies. 'Then jump aboard.' 'Where do you put in?' asks Dionysus, 'The Ravens? Is that a stop?' 'Yes, by god,' says Charon, 'a special stop just for you. Get in.' Charon lets Dionysus himself row, in time with the croaking of the frogs, which gets steadily faster. But he does have to pay his two obols.

In the fourth century BCE Plato tells in his *Phaedo* a rather sinister tale of the course of the subterranean rivers. He does not locate them in a particular region, but one thing is clear: the Acheron

> passes under the earth through desert places, into the Acherusian Lake: this is the lake to the shores of which the souls of the many go when they are dead . . . And those who appear to have lived neither well nor ill, go to the river Acheron, and mount such conveyances as they can get, and are carried in them to the lake, and there they dwell and are purified of their evil deeds.

The Acheron flows in the opposite direction to the Oceanos, while the Styx is the same river as the Cocytus and flows into the Tartarus. So, too, does the glowing hot Pyriphlegethon, but in the opposite direction. Plato does not mention Charon.

Greek writers in Roman times continued to consider the Acheron the most important river of the Underworld. In the second century BCE Pausanias says in his unsurpassed *Description of Greece*, also known as the 'first *Baedeker*':

> Among the sights of Thesprotía are a sanctuary of Zeus at Dodona and an oak sacred to the god. Near Cichyrus is a lake called Acherusia, and a river called Acheron. There is also Cocytus, a most unlovely stream. I believe it was because Homer had seen these places that he made bold to describe in his poems the regions of Hades, and gave to the rivers there the names of those in Thesprotía. (I:17.5)

Pausanias believes that the existing Acheron in Thesprotía is the same as the mythological Acheron. This is clear from his description of a painting by the artist Polygnotus of Odysseus' journey to the Underworld: 'In the painting is a river, which is obviously Acheron . . . And there is a boat on the river, and a ferryman with his oars.' He later refers to the ferryman as Charon. The Greeks therefore clearly see the Acheron as the river you have to cross to get to the Underworld.

So what is the Styx to the Greeks? There is only one modern Styx, and on the map it lies 300 kilometres from Thesprotía in ancient Arcadia (now Achaia) on the northern side of the Peloponnesus. Hesiod, Homer's contemporary from the eighth century BCE, relates in his turbulent genealogy of the gods, *Theogony*:

> the famous cold water which trickles down from a high and beetling rock. Far under the wide-pathed earth a branch of

> Oceanus flows through the dark night out of the holy stream, and a tenth part of his water is allotted to her [the Styx]. With nine silver-swirling streams he winds about the earth and the sea's wide back, and then falls into the main (24); but the tenth flows out from a rock. (786–92)

The Styx therefore flows out from a cliff and partially underground until it reaches the sea. Our unfaltering fifth-century BCE historian Herodotus writes:

> For the waters of the Styx, as the Arcadians say, are in that city [Nonacris], and this is the appearance they present; you see a little water, dripping from a rock into a basin, which is fenced around by a low wall. Nonacris, where this fountain is to be seen, is a city of Arcadia near Pheneos. (*The History*, VI:74.1)

It's true! Pheneos lies some 10 kilometres southeast of the Styx.

The Greek geographer Strabo says the same thing four centuries later: 'And near Pheneos is also the water of the Styx, a small stream of deadly water which is held to be sacred.' In a story about Heracles' journey to the Underworld, Homer calls the waters of the Styx 'steep', while Pausanias uses similar words in his travel guide eight centuries later. None of the Greeks speak of Charon crossing the Styx in his

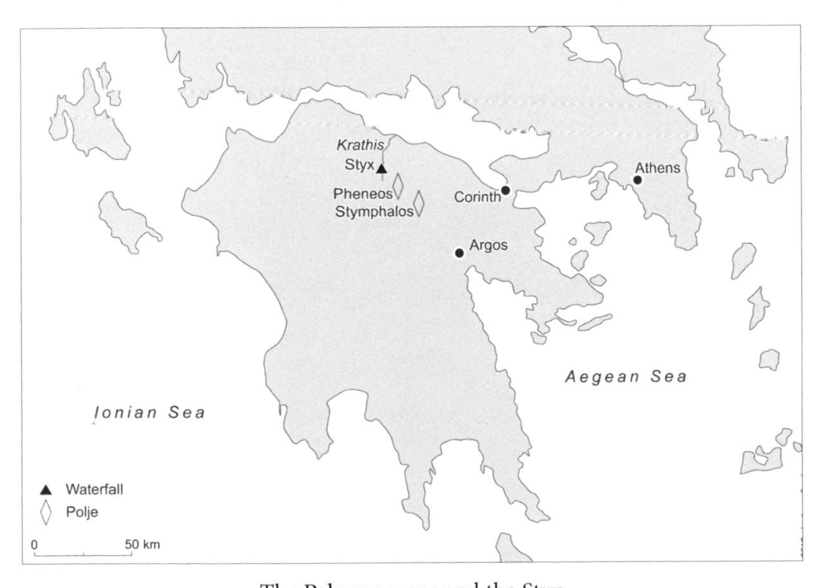

The Peloponnesus and the Styx.

boat, since that would be difficult over a steep waterfall; for them, he is the ferryman across the Acheron. You can swear oaths on the Styx, and bathe Achilles in it, but you don't sail on it in a boat.

With the Roman writers, however, the names of the rivers start to change and get mixed up. This appears to be Virgil's fault. When Sybil tells Aeneas how he has to get to the Underworld, she says that he has to cross the Styx twice, but once he has entered through the foul jaws of stinking Avernus, Charon will be waiting for him on the bank of the Acheron, which 'spews all its sands into Cocytus', while in reality the Cocytus is a tributary of the Acheron. It becomes even more confusing when Sybil says to Aeneas just before he is carried across by Charon: 'You see the deep pools of Cocytus, on the Styx', while Charon catches sight of them 'from the Stygian wave'.

Anyone who studies the attempts of modern writers to chart the courses of Virgil's subterranean rivers in the *Aeneid* will discover that they too have become completely confused. No one dares any longer to say in which direction the rivers flow, or which river is a tributary of the other. In one reconstruction the Styx and the Pyriphlegethon even cross each other.

Other Romans are equally confused. In his *Metamorphoses* Ovid – a contemporary of Virgil – describes the Styx no longer as a waterfall, but a black swamp, a misty pool, over which Charon refuses to carry Orpheus for the second time, because he has not succeeded in saving Eurydice. In his first-century play *Hercules Furens* Seneca reverses the roles: the Styx is a quiet river and the Acheron a mighty roaring torrent through which boulders roll. In Seneca's account Heracles overpowers Charon, forcing him to take him over the Styx.

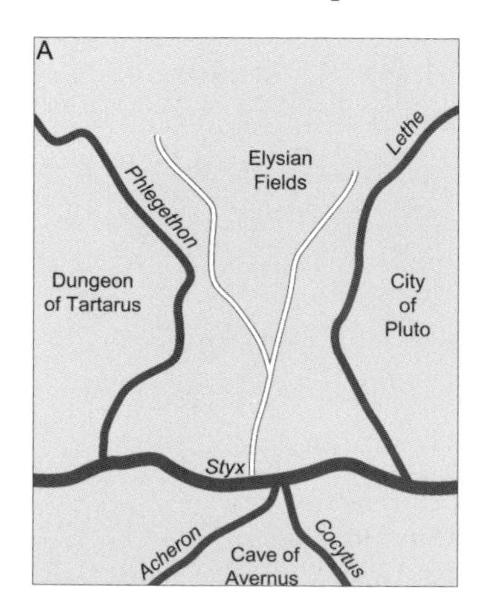

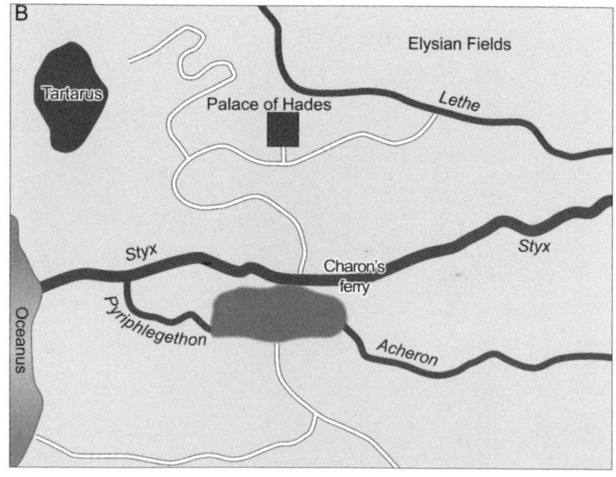

I understand the problem; Virgil's story requires him to situate the entrance to the Underworld in Italy, near Lago Averno, making it illogical to leave the subterranean rivers in Greece. He retains the Greek names, but they have lost their geographical context and become purely allegorical. That had far-reaching consequences for later authors. Pliny the Elder and Strabo identify what is now Lago Fusaro, at Cumae in Italy, as the Greek Acherusian Lake. Strabo also recounts that, according to legend, the mysterious Cimmerians lived around Lago Averno. These are the people that, according to Herodotus, lived in the Crimea and gave their name to the peninsula. 'Those who migrate back and forth' had apparently followed the story to Italy.

This led in the nineteenth century to ethnographer Andrea De Jorio drawing a map of Aeneas' journey in which he gave all the lakes and rivers around Lago Averno Greek names: he not only called Lago Fusaro the Acherusian Lake, but also identified two nearby marshes as Cocytus and Lethe. Lago Lucrino, which connected Lago Averno to the sea via the famous tunnel, was now the Styx. I had my suspicions when I saw it lying under water.

Virgil probably even misled travel guide Pausanias, who lived two centuries after him. Pausanias writes in his *Description of Greece* that after the death of Eurydice, Orpheus travelled to the Oracle of the Dead at Aornos in Thesprotía. Strabo, too, says that outside Ephyra, to the north, 'is Aornum, with a sacred cave, which is called Charonium, since it emits deadly vapours'. Aornos! Deadly vapours that cause birds to fall dead from the sky! But there are no deadly vapours in Thesprotía. Aornos is the name that Virgil gave to Lago Averno. There is an oracle

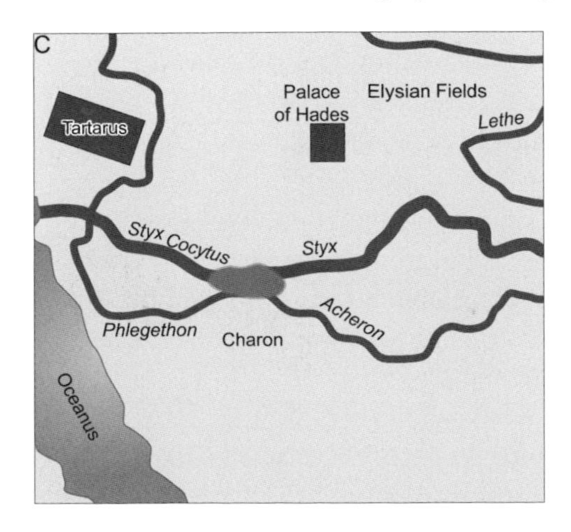

Three reconstructions of the underground rivers, based on Virgil.

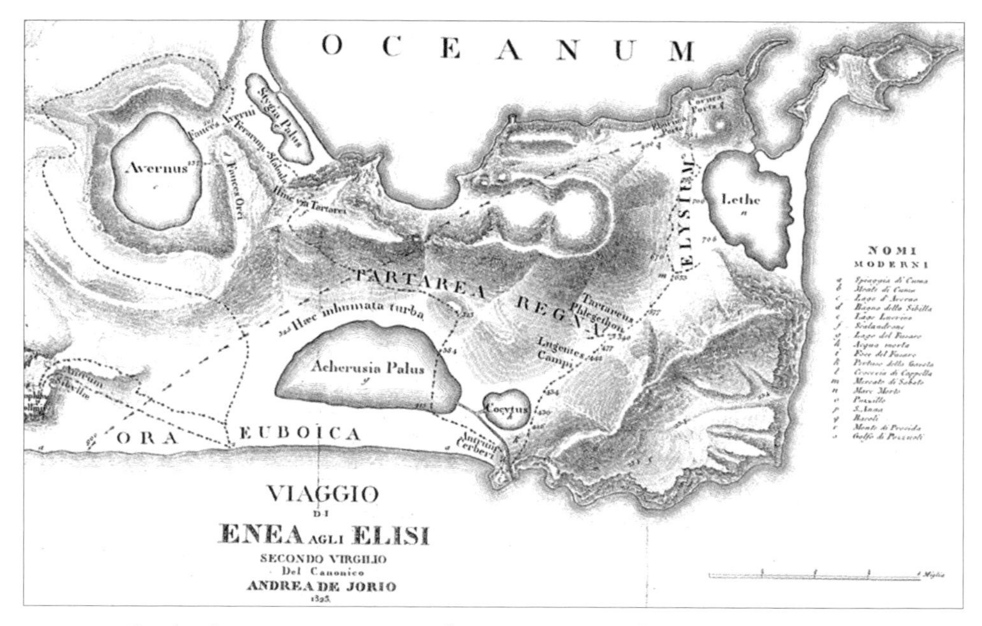

Greek subterranean rivers around Lago Averno in Italy; Virgil's Underworld as drawn by Andrea De Jorio.

in Thesprotía, the aforementioned Necromanteion, but it does not exude vapours. And there is not a single volcano to be found in the whole region. In this way Greek rivers have migrated to Italy and Italian birdless lakes to Greece.

Ten centuries later, Virgil gets the opportunity to set the record straight. As he leads Dante through the Inferno, he tells all about the subterranean rivers. The story begins with Virgil and Dante arriving at a place where a blood-red stream flows out of the forest. Virgil explains:

> Every part of him [Mount Ida, on Crete] . . . is broken
> by a fissure dripping tears down to his feet,
> where they collect to erode the cavern's rock;
>
> from stone to stone they drain down here, becoming
> rivers: the Acheron, Styx and Phlegethon,
> then overflow down through this tight canal
>
> until they fall to where all falling ends:
> they form Cocytus. What that pool is like
> I need not tell you. You will see, yourself. (XIV:113–20)

So, they rise on Mount Ida in Crete, divide, go underground at some point, and come together again in the Cocytus, which, as we shall see, is in the deepest part of hell. A completely illogical explanation, you might say, as rivers don't divide and reunite in this fashion, nor does it bear any resemblance to what the Greek writers describe. Dante should actually be ashamed of himself, as someone who understands perfectly how rivers behave. Anyone who writes: 'As at Arles, where the Rhône turns to stagnant waters', is aware that the drop of the Rhône is so small at Arles that it has to divide into separate arms, forming the start of the Rhône delta.

We will blame Virgil himself for the confusion. It is his fault that, in modern times, the Styx has replaced the Acheron as being practically synonymous with the 'Underworld'.

The question remains as to whether the Acheron and the Styx really do flow underground. I stare at my maps and at satellite images, but to no avail. There's nothing else for it; I have to go and see for myself.

Charon is a suntanned young man. He rents out small boats shaped like slippers, with a flat bottom and bulging round sides, like floating pizzas that can hold about six people. His company is called Acheron Rafting and there is an endless line of tourists waiting on the bank of the Acheron until it is their turn to spend half an hour paddling

The Phlegethon, the River of Blood: a stream of liquid red sulfur, Tengiz, Kazakhstan.

helplessly downstream to the bridge. The oars are purely to make sure you don't collide with the bank; the river does the rest.

Their cars are parked higgledy-piggledy, huddled around a dusty plane tree to make sure that when their owners come back, they are still in the shade. Between the cars are stalls selling soft drinks, *shashlik* and honey. The honey-seller even has a small rack of books for sale. They are all locally published and written by the same author, Spýros Mouselímis. On the dust-spattered cover of *The Ancient Underworld and the Oracle for Necromancy at Ephyra* his name is spelt wrongly as Sryros Moyselimis, an endearing overcompensation in the transliteration from Greek to English: the Greek Π is our P, but the Greek P is our R, which explains the 'Sryros'. The OY is not completely correct either, but Spýros himself could not have seen the error: the English translation of his Greek original was published in 1989, five years after his death. Spýros was a self-made historian, and the book is a collection of articles he wrote in local newspapers. I love books like this; they contain details that you can't find anywhere else, and they are bursting with a sort of regional chauvinism. I am itching to read it, but first I want to make the boat trip myself.

The site where Charon has his company is called the 'Springs of Acheron'. There is a campsite, a pony club and a bar, the Paradeisiakós. The springs of the Acheron are not here at all – they are further upstream – but it sounds good and there are signs urging you to visit the place for kilometres around. This is the place where the Acheron emerges from a deep gorge and, delighted at regaining its freedom, spreads out on the coastal plain.

But I don't want to paddle downstream; I want to go upstream, into the gorge. That's where it is most likely that the Acheron flows underground, so that it can at least live up to its reputation as a subterranean river. I ask Charon whether he can also take me upstream through the gorge, or downstream from the other end, where it starts. He bursts out laughing. It would take him six hours to get the boat to the start of the gorge if I wanted to come downstream, and upstream is impossible because the current is too strong. 'If you want to go upstream, you'll have to wade through the water, and swim when it gets too deep', he says.

OK, wading and swimming it will have to be. I don't get discouraged that easily. The gorge is awe-inspiring. To the left and right are 50-metre-high walls of solid, bright white Jurassic limestone: the Pantokrator formation, I know from the geological map. Loose blocks of rock hang above my head, meagre trees cling desperately to the smallest bumps in the cliff, and water gurgles into the river from large, dark cracks. Dante's words come to mind:

> The river's bed and banks were made of stone
> so were the tops on both its sides and then
> I understood this was our way across. (*Inferno*, XIV:82–4)

But it is not deep: the water comes up to my knees. It is crystal clear and the bed is strewn with equally bright white limestone pebbles, interspersed with a few of dark green flysch sandstone, washed down from upstream. Small bubbles of carbon dioxide gas rise up from between the stones. I swim through it. The current is strong, but the gorge soon widens and I can wade again. Then I have to swim again, through a second narrow section where the river has violently crushed broken-off tree trunks against a fallen boulder. But above my head the sky continues to tease me. The steep rock walls never come together completely to form a cave. Nowhere are there signs of a collapsed roof. There is nothing to suggest that this is the Underworld. Everything is white and clear.

Then there is a confluence, and such deep potholes in the limestone that I have to cross the river by an old stone bridge near the ruins of a watermill. Further upstream the solid limestone gives way to thin layers of Cretaceous limestone containing banks of black flint, de Vigla formation. The banks of the river are less steep and are covered with beech and plane trees; the bed becomes wider and, here and there, even clayish. The river will certainly not go underground here, as that does not happen in this kind of limestone. I continue to a large bend and then turn back, disappointed that I still haven't seen a subterranean river.

But perhaps on the other, upstream side of the gorge? There's no need to wade: there is a road leading up there, with a sign pointing to the Gates of Hades. That's clearly where I need to be. Now it is the Cretaceous limestone that creates a gorge and it is a beautiful place, but there are no signs of an underground river here, either. It would actually not be logical to find the entrance to the Underworld here. After all, why would you bring the bodies of the deceased such a long, roundabout way upstream just to send them back down the river again?

There is one small comfort: the layers in the rock lean upstream. It is a detail that many non-geologists would not notice. If they lean upstream, as they do here, it means that as I follow the river downstream,

The Gates of Hades: the Acheron flows into the gorge from below left to top right; the layers of Cretaceous limestone dip towards the viewer, against the current.

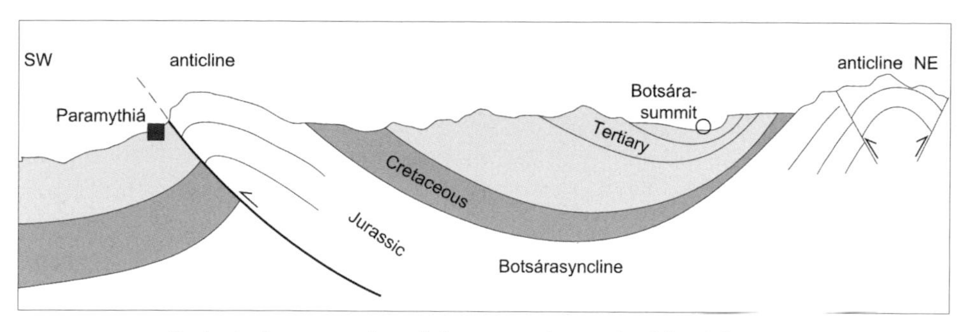

Geological cross-section of the area to the north of the Acheron at Paramythiá, Thesprotía.

the layers become increasingly old. I am in the core of a syncline, a concave, hollow fold, with the youngest rock at the centre. Anticlines are the opposite: convex folds with oldest rocks at the core.

The figure above shows a cross-section of the mountain range through which the Acheron flows, though a little further to the north, near to the town of Paramythiá. In the centre is the Botsára syncline, with small anticlines visible to the left and right. The Cretaceous limestone and younger layers are dark grey and grey in the drawing, while the older Jurassic limestone is white. If you imagine that the end of the gorge is at the core of the Botsára syncline, the Botsára peak, and you move to the left – that is, to the southwest (downstream) – you will come cross increasingly older layers. In effect you are going deeper into the Earth's crust, deeper into the Underworld. Although the river does not flow underground, it does flow from the Gates of Hades deeper into the Earth's crust, thanks to the plate tectonics that have caused the fold.

I also have to go to the mouth of the Acheron since, according to Homer, Odysseus approached the Acheron from Oceanus, from the sea. Yet the historian Thucydides said in the fifth century BCE that the Acheron does not flow directly into the sea at all, but through the Acherusian Lake. 'By this city the Acherusian lake pours its waters into the sea. It gets its name from the river Acheron, which falls into the lake', he wrote. Plato, and much later Strabo, Pliny and Pausanias, also mention the lake. Perhaps Charon had to carry you across the lake first before you got to the Acheron. That is more logical for a ferryman than a river that you can simply wade through, and on which you can only sail up or down. Time to find out.

Now I drive from the point where the river emerges from the gorge, at Acheron Rafting, downstream through wide, low-lying land, until

I finally reach a round bay, Glykýs Limén, the Port of Sweet Waters. Strange, I haven't seen any sign of a lake, and surely the classical writers were well aware of the difference between a saltwater lake and a sea inlet? The Acheron simply flows straight into the sea on the southern side of the bay, separated from it only by a pier of large blocks of rough limestone.

On the narrow beach of the coastal town of Ammoudiá stand dozens of caravans and campervans from all over Europe. The season is over, and it is largely pensioners who spend the winter here, secluded in their mobile homes. The hotels are empty; the small boats moored along the banks of the Acheron lie idly waiting for customers. What do they offer? 'Boat trip to the Gates of Hades'. Again, I am dumbstruck. I've just been there, a long way upstream, and you can't get there at all by boat: the bronzed Charon had assured me of that. So where do they go? Another Charon, this one with a white baseball cap and an open red-checked shirt, points seawards: the almost circular bay is enclosed by two rocky capes, leaving only a narrow opening to the sea. They are the Gates of Hades, he tells me. Is that the rock that Homer was referring to? Is the Acherusian Lake more than fiction? Surely a man like Thucydides would not just make that up?

The Glykýs Limén, the Port of Sweet Waters, with the mouth of the Acheron to the right. Not a lake in sight, but Homer's description of the bay of the Laestrygonians could just as easily refer to this location.

The circular beach ridges show that Glykýs Limén must have been much larger in the past and has since silted up.

American geographer Mark Besonen of the University of Minnesota devoted his thesis to this question. He studied all the classical sources, hunted down old maps, made 28 boreholes in the lowlands of the Acheron, counted shells, pollen grains and other microfossils, conducted carbon-14 dating, and recorded his results in eight palaeogeographical maps. He proposes that, since Homer doesn't mention it, the Acherusian Lake had not yet been formed when Odysseus landed. But, as we know, absence of evidence is not necessarily evidence of absence. Besonen balances as much on the edge between science and myth as I do. The first thing he concludes is that Glykýs Limén must have been much larger in ancient times. And you can see that in the landscape: inland from the bay there is a magnificent concentric complex of circular beach ridges that precisely follow the contours of the present-day bay. The further inland you go, the larger and older the ridges become, showing that the bay is gradually becoming smaller. The whole area is still very swampy, but is far from the current course of the Acheron, making it an unlikely candidate for the Acherusian Lake. Besonen also uses historical arguments: at least three times in the past, the bay was used as a port of refuge for a fleet of hundreds of warships. The current bay could never have accommodated them, so it must at some time have been larger. But I have my doubts about the force of that argument: you need only to see how many hundreds of campers fit along the shores of the bay these days.

The Acherusian Lake was clearly here, Besonen writes, and the reason for its disappearance is prosaic: the final swampy remains were reclaimed in the First World War for agriculture. In his reconstruction, the Acheron flowed directly into the much-larger bay 4,000 years ago. Its delta lay 6 kilometres to the east of today's coast, at a constriction in the lowlands between the villages of Tsouknída and Mesopotamon. Later, after Homer's time but before Thucydides – that is, between the eighth and the fifth centuries BCE – the constriction silted up, creating the Acherusian Lake. Through the continuing accumulation of silt, it became increasingly shallow and the bay gradually filled up. But you would certainly have needed to make use of Charon's services to sail on the Acheron until at least the Middle Ages.

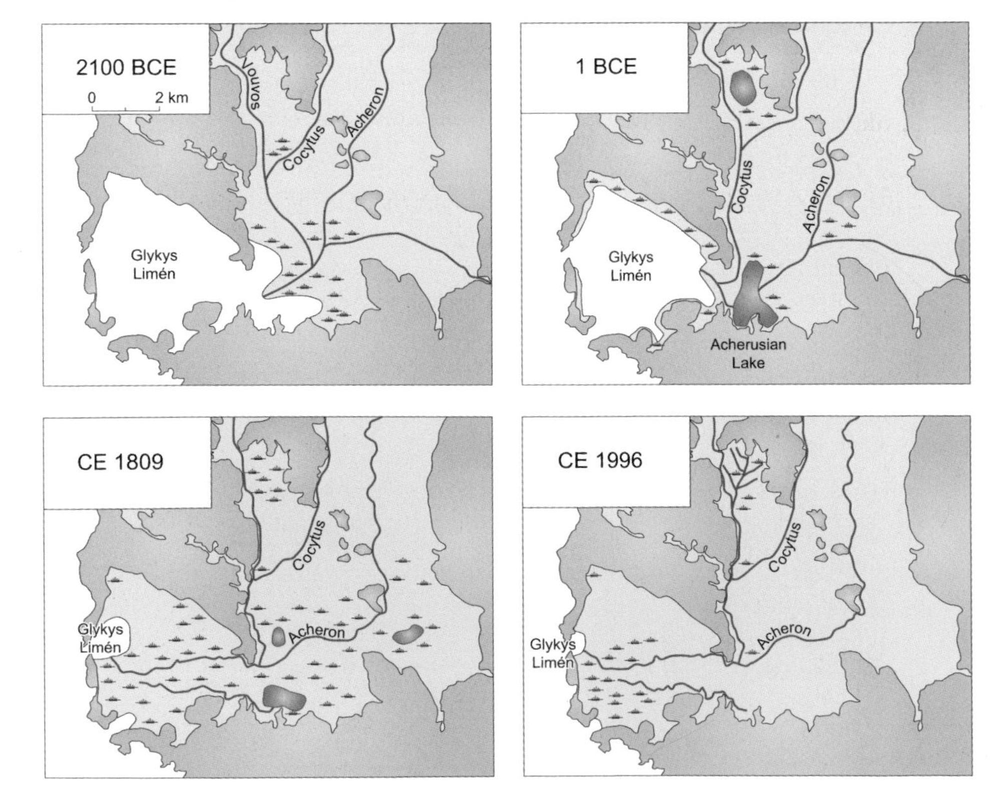

Four stages in the history of the Acherusian Lake.

Four thousand years is a long time and the appearance of the landscape can change dramatically, just as at Lago Averno. Without geological history, you cannot reconstruct the history of humankind and of its legacy of thought.

Spýros Mouselímis, the priceless author of the dust-spattered book on the stall at Acheron Rafting, writes that in the spring you used to be able to hear a deep lowing coming from below the ground in the Acherusian swamps, as though a bull had been locked up in a dungeon. He heard it himself when he was a teacher in Tsouknída. It had been proven that the roar was made by the flowing of subterranean waters, he wrote, and that no one had ever heard it again after the swamp had been drained. It was that sound that led people to believe that Hades was here under the ground.

Mouselímis was also the man behind the discovery of the aforementioned oracle of Necromanteion, the Aornos of Thesprotía. Another Gate of Hades! The Underworld is starting to look like a rabbit warren. It is an impressive complex on top of a limestone hill in the middle of the plain, where the Acheron and the Cocytus come together. The outer walls, a good metre thick, are constructed of enormous, irregular blocks of limestone, so shaped that they fit together exactly, just like those in the famous Inca constructions at Sacsayhuamán. Like everything else in this region, it is now largely a ruin, on top of which seventeenth-century Christians unfortunately considered it necessary to build a monastery to John the Baptist, as if to smother all thoughts of heathen gods.

Before the 1950s it was not even possible to see what lay beneath the monastery. Mouselímis wandered around here for many years, dreaming of setting up excavations, because he thought he would find a royal tomb. He approached one archaeologist after the other and, in 1958, succeeded in persuading Sotírios Dákaris, professor of archaeology at the University of Ioannina, that it was a good idea. The results were spectacular: instead of a royal tomb, they found a sacred chamber, an oracle, surrounded by six rooms containing large pottery urns of wheat, barley, chick peas, honey, clay dolls, carving tools, pieces of iron that were used as coins, oil lamps and pieces of chalk. Everything had been preserved when the oracle collapsed, after the Romans set it on fire. And they found iron tools, nails, a marble pedestal, a small copper pot, two flint balls from enemy catapults, spears, and a large quantity of plates and flasks. There were two rooms full of 'burned sulfur' and, in another, a chunk of pure sulfur. Perhaps that was why birds fell dead from the air.

The underground vault of the Necromanteion.

But the greatest discovery was even deeper underground: a completely intact vault, hacked out of the rock, which was christened the House of Hades. It looks to me like the same limestone from which the oracle was built, and was therefore an underground quarry. Some cynical modern scientists allege that the vault is a cistern, an underground reservoir for drinking water, and that the oracle is in fact a rich, fortified farmhouse from the third century BCE, too young to have served as an oracle for Odysseus, but I wouldn't give any credence to such rumours.

We still don't know how Acheron and the Acherusian Lake found their way into the Underworld. But we'll leave the question unanswered for the time being, because I first want to go to the Styx. That might clear a few things up. Time to go to Arcadia. But there, too, the Styx proves to strike just as little fear into people's hearts as the Acheron – quite the contrary. This mountainous area is very popular with Greek skiers, and the highest point of the piste is called Stygós, after the Styx. Skiing on the Styx! That's something I would never have thought of. But now, in September, there is no snow. I follow a clearly laid out trekkers' path, the Tessera Elata, to the Ydata Stygós, the Waters of the Styx. It starts at a height of 1,000 metres near the village of Mesorougi, but to get to that point you have to negotiate nine hairpin bends. Perhaps that was what Hesiod meant by his 'nine swirling streams'.

I examine the rocks alongside the path closely. At first they are phyllite, slate-like rock with a pretty satin sheen, and twinkle in the sunlight. They form the impermeable basement of the Peloponnesus. Small streams flow out of a walnut orchard and cross the path. Then, at 1,300 metres, the limestone begins and the water disappears. I look up and see the imposing sheer rock face of the Khelmos – 1,000 metres of horizontally layered, bright white Cretaceous limestone. I still have a long way to go. The path climbs steeply, sometimes through forest, and then through bone-dry gorges cut into the limestone. They are marked on my map as the First, Second and Third Tabouri, and there are more. They are deep ravines, full of enormous limestone boulders and rubble, and without a drop of water. Everything is as dry as a bone. I am surprised: how can the waters of the Styx possibly be here? The path continues, and is sometimes so narrow that you have to hold on to steel cables bolted to the rock. There are gigantic screes with boulders the size of houses. Enormous masses of rock must once have tumbled down the unstable slopes. Here and there it is still green, and at the bottom of the valley I can see potholes that must have been created by running water, but now, in September, everything is completely dry. I start to think I am on a wild goose chase.

And then suddenly, as I turn a corner, at a height of about 1,700 metres, I see it: the beautiful ribbon of a waterfall plunging down from a high ridge, silver drops glistening in the late sunlight. But there is so little water that it turns to spray before it can reach the ground – not even enough to wet Achilles' heel.

And it is not at all black. The water is crystal clear, though it has left a dark vertical stripe of moisture and moss on the horizontal white layers of limestone. That is the only reason the mountain above is known as Mavronerí, Black Water. Next to the black stripe of the Styx there is a second, reddish band of the same width, probably where the river used to fall over the edge. Perhaps it still does in the winter, when there is more rainfall.

It is an astounding waterfall. The water plunges abruptly downwards over the edge of the ridge. Right next to it is a dry ravine, but the Styx has chosen not to make use of it. And it is apparently still too young to have created its own ravine. But the most surprising thing is that it is there at all – that, in the midst of all that bone-dry limestone, where every shower of rain disappears immediately into the cracks of the porous rock, there is suddenly a waterfall. Perhaps it is because of that miracle that the Styx has acquired such a special place in mythology.

And then I suddenly realize just how wonderfully accurate the old descriptions of Hesiod, Herodotus and Pausanias really are. Of

The real Styx.

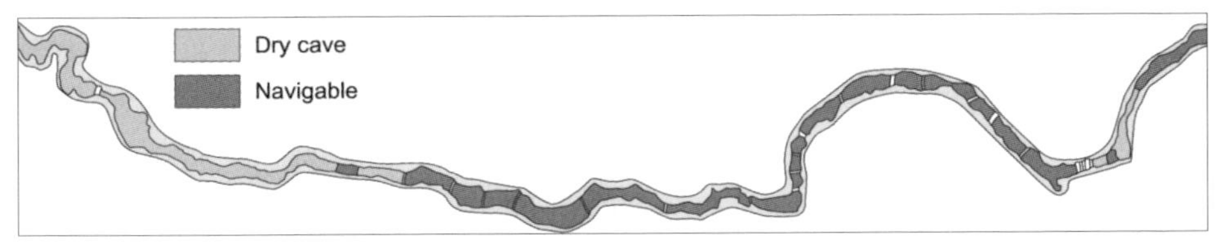

The Spílaio ton Limnón, clearly the course of an old subterranean river.

Joachim Patenier, *Charon Crossing the Styx*, *c.* 1520.

course you can't sail on the Styx. Charon has no business here. The longer I look at it, the stranger it seems that this waterfall, the only above-ground water in the whole region, should become the major symbol of the river of the Underworld.

Did the ancient Greeks not know of the existence of subterranean rivers? Of course they did. Five kilometres to the southwest of the Styx there is a cave, the Spílaio ton Limnón, the Cave of Lakes, which was inhabited as early as the Stone Age. The predecessors of the ancient Greeks used the cave for storage. Tools and 30 different kinds of vases have been found there, together with the skeletons of thirteen young girls. It is a very high dripstone cave in perpendicular limestone layers, with fantastic stalactites and draperies on the ceiling and grotesque, mammoth-like stalagmites on the ground. It appears to follow the course of an underground river, which it once was. Now it is no longer connected to the current river system, and is 80 metres under the ground, but in the winter it is full of water and you can go down it by boat. The ancient Greeks probably only knew about the first part; the rest was not discovered until the twentieth century.

This cave is not unique. There are many more in Greece, all old river courses, sometimes stacked on top of each other and with various arms. The Greek speleological association has inspected some 7,000 of them, writes Anna Petrochilou in her book *Les Grottes de Grèce*. Of course the ancient Greeks knew about them: their myths are full of stories about caves.

Herodotus himself describes how a river disappears under the ground and reappears elsewhere: 'This stream is reported to flow from the Stymphalian lake, the waters of which empty themselves into a pitch-dark chasm, and then (as they say) reappear in Argos, where the Argives call then the Erasinus [Erasinós].' Ovid, Strabo and Pausanias also mention this lake. The nineteenth-century explorer William Marin Leake believed that they would have discovered that the water from the Erasinós came from this lake by throwing floating objects into the water: the Greeks often cast honey cookies, garlands of flowers, human hair and other objects into the water as offerings to the water gods, and they probably reappeared in the Erasinós, 40 kilometres to the south. Leake's contemporaries did the same with pine cones, which also turned up in the river.

Another nine hairpin bends and four mountain passes later and I'm there. The steep, rocky river valley with scraggy trees gradually widens and gives way to the neat, rectangular cornfields of the Stymphalian Lake plain. I wonder if this really is it, as I have already passed a dry lake plain, a little earlier, at Pheneos. But in the middle of the plain there is still water, a shallow pond surrounded by reeds. There is nothing to suggest that it is a legendary place but here lived the Stymphalian birds, terrifying man-eating creatures with bronze beaks and steel wings. They could bite through helmets and armour. Fortunately Heracles defeated them as one of his heroic labours. It is a nature reserve, part of the Natura 2000 network, and there is a wonderful modern ecological museum high up on the bank. I look through the rotating binoculars on the terrace of the museum, searching for the Stymphalian birds, but don't see a single one. In fact I don't see any birds at all. Heracles did his job well – perhaps too well.

After Heracles' labour, Emperor Hadrian built an aqueduct from the Stymphalian Lake to Corinth to irrigate the fields near the sea, a

A boat on the underground river in the Spílaio ton Limnón, the Cave of Lakes, 5 km from the Styx.

distance of 70 kilometres. Quite a feat of engineering. The remains can still be seen here and there. But that's not what I came here to see. I wanted to find out more about that subterranean river. There must be a place around here somewhere where the water goes underground, as Herodotus – and many other classical authors after him – put it. A *katavóthra*, as it is called in Greek, or a ponor, to give it its scientific name: a hole where a river goes underground only to resurface later – in this case, 40 kilometres to the south. I can't think of anything more appropriate than a ponor being the entrance to the Underworld.

The Stymphalian ponor is famed. According to the Greek geographer Strabo, in the fourth century BCE, the Athenian general Iphicrates tried to seize the city of Stymphalos by blocking the ponor with a large quantity of sponges so that the city would be flooded. He only changed his mind after being warned by Zeus. And Pausanias tells the story of a tree trunk getting stuck in the hole, so that the water could no longer flow out of the lake and the water level rose substantially. At the same moment – at least, American 'hydromythologist' Cindy Clendenon thinks that it was at the same moment, but Pausanias is not entirely clear about it – there was a hunter who was stalking a stag, and when the stag leaped into the water, he jumped in after it. In the meantime, however, the tree trunk had worked loose and the water could once again flow down through the ponor. The enormous suction created a whirlpool that dragged the hunter and the stag through the hole and down into the Underworld.

Now that's what I came to see. And it's there, too. On the southeast side of the lake is a semicircular cliff in the limestone. A quiet stream runs towards it and disappears into what must be the ponor. Unfortunately the actual hole is hidden from view by a grid with closely spaced vertical bars that resemble the baleen of a whale. That at least means that no one can be sucked into the Underworld.

Subterranean rivers are not unusual. They occur all over the world, as long as there is limestone in the ground. They can sometimes also be found in gypsum, salt or even ice. It's necessary to give a little chemistry to explain how they come about. Just grit your teeth for a few sentences. It's actually part of the whole carbon cycle that secretly comes to the surface in this book now and again, like little bubbles rising up through the water. Limestone consists of calcium carbonate, $CaCO_3$. Unlike most other rocks, limestone dissolves relatively easily in water – not that easily in pure water, but certainly in rainwater. That is because rainwater always contains a little CO_2, which it has absorbed from the atmosphere. That makes it slightly acidic. Water is H_2O, two hydrogen atoms and one oxygen atom. A CO_2 molecule can bind itself to one hydrogen atom and one oxygen atom from the water, creating HCO_3^-, the bicarbonate ion (it is an ion because it has a negative electrical charge; that is what the minus sign refers to). The other, leftover H^+ ion remains free, and the concentration of these ions determines the acidity of the water. If the acidified rainwater comes into contact with limestone, the bicarbonate ions meet their fellow carbonates since, as we have seen, limestone is calcium carbonate. The hydrogen ions in the rainwater can extract the carbonate ions from the limestone, so that they form HCO_3^-, the same bicarbonate ion that we saw in the rainwater. And that ion can dissolve very easily, as can the remaining calcium ion, so that the limestone dissolves completely: $CaCO_3 + H_2O + CO_2 = Ca^{2+} + 2HCO_3^-$. Both ions disappear with the groundwater into the rivers, and ultimately end up in the sea. What happens to them in the sea we shall see later.

Limestone does not dissolve simply from top to bottom: the rainwater first concentrates itself in the porous ground. Because of the humus, the ground often contains much higher concentrations of CO_2 than are present in the air, which speeds up the dissolution of the

The Stymphalian ponor.

underlying limestone. The acidified water finds its way downward through cracks in the underlying rock, dissolving more limestone as it goes and thereby making the cracks steadily wider. If the dissolution process continues underground, whole cave systems can be created, with fantastic dripstone formations, with stalactites and stalagmites. There is less CO_2 in the air in the caves than in the ground, causing the limestone to crystallize. Each droplet on a stalactite forms a gossamer-thin calcite crystal, and over thousands of years, an icicle of limestone slowly grows downward until stalactite and stalagmite meet and form a column.

I walk through the Pérama Cave, another magnificent river-shaped dripstone cave near Ioannina, the capital of Epirus, in northern Greece, 30 kilometres to the north of the Acheron gorge. The cave was discovered in the Second World War by a schoolboy looking for somewhere to hide from the heavy bombing. Suddenly I notice the floor of the path leading through the cave, and I am astounded: it is a whole mosaic of sawn-through gobstoppers. Then I realize what they are: stalagmites that have been cut through to make a path through the cave. But why do they have the same concentric patterns? Why is the limestone deposited on the stalagmite sometimes white, sometimes yellow and sometimes grey?

That is because the CO_2 content in the atmosphere and in the rainwater depends on the temperature: if it gets colder, more dissolves and if it is warmer, more crystallizes. Together with minor forms of contamination, that creates the colour differences: these are speleothems, cyclical cave deposits. The proportion of oxygen isotopes in the deposited limestone also shows that the deposits have varied. If you very carefully drill out each layer of the stalagmite separately with a dentist's drill and analyse them, you will obtain an excellent record

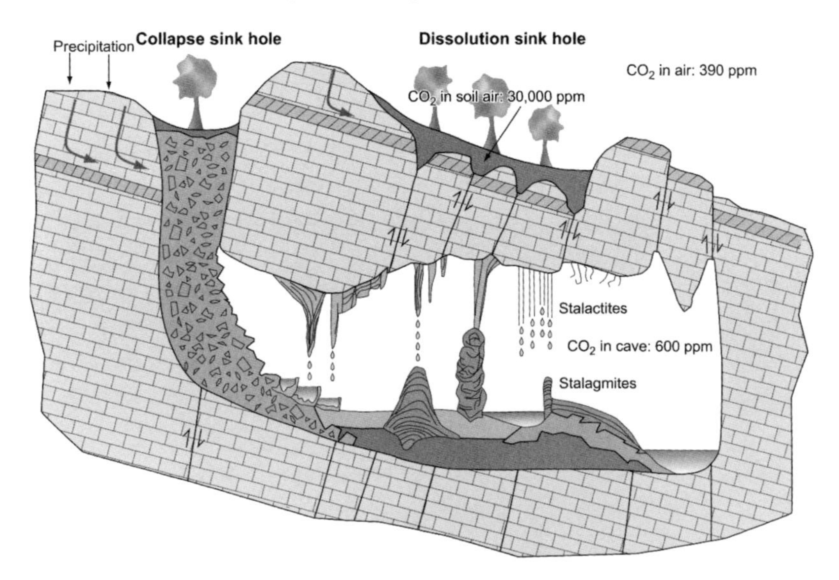

Formation of stalactites and stalagmites in a cave, and the role of CO_2.

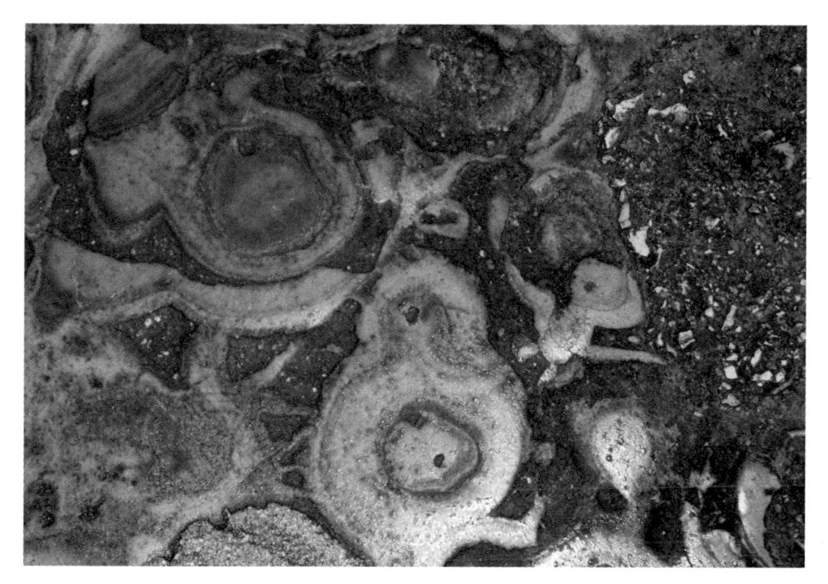

Mosaic of gobstoppers: sawn-through stalagmites, Pérama Cave, Greece.

of climate changes on the Earth. Speleothems in the Soreq and Peqiin caves in Israel, formed in the same limestone from which Jerusalem was built, record the last two ice ages, from 250,000 years ago to the present. They show that it was much drier in the ice ages in the eastern Mediterranean region than in warm periods like now. In this way stalagmites provide a unique illustration of our climate.

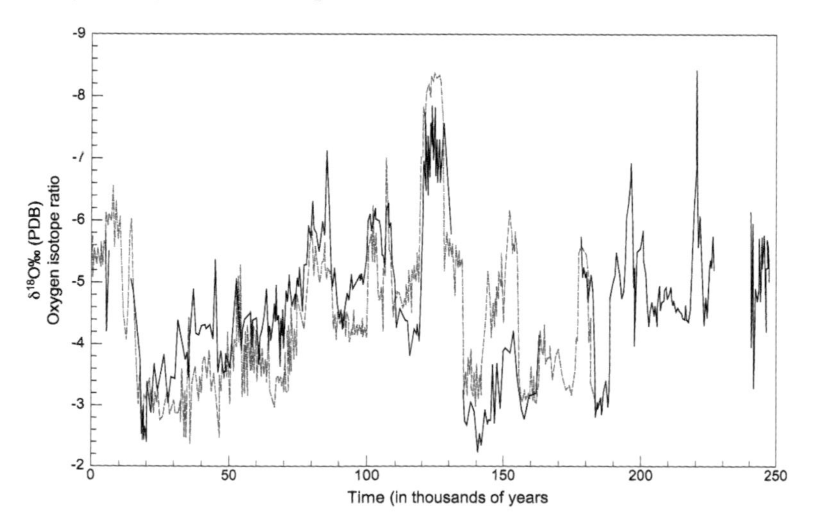

Climate registration of two ice age cycles in the Soreq Cave (dotted line) and the Peqiin Cave (solid line) in Israel: the peaks show warm periods and the troughs are ice ages.

Nearly all caves are part of a subterranean river. The river dissolves the limestone from its bed, and if that continues for long enough, it forms a ponor and a considerable stretch of the river can disappear underground. Above ground, this leaves a dry valley. Some subterranean rivers even flow into the sea under water. In caves like the Pérama and the Spílaio ton Limnón, contact with the original river has been lost. That happens if the mountains have been pushed further upwards. Sometimes you find four levels on top of each other: each time the mountains rise slightly, the river sinks down another level. But sometimes the river water finds a shorter way to the sea and leaves the cave entirely.

If so much limestone dissolves underground that it can no longer support the rock above it, the ground can collapse. This creates sinkholes, which are often in the most unexpected places. Sometimes the entire 'roof' of an underground river can collapse, leaving collapse valleys like the Stymphalian Lake and the adjacent Plain of Pheneos. The Serb pioneer Jovan Cvijić called these valleys *polje*, a Slovenian word meaning field, which is now used internationally. All phenomena in the landscape that are related to the dissolution of limestone are

Sinkhole (doline, or cenote) on an intersection in Guatemala City, 20 m across, 30 m deep, 2 June 2010. A three-storey clothing factory disappeared into the hole, fortunately three hours after closing time.

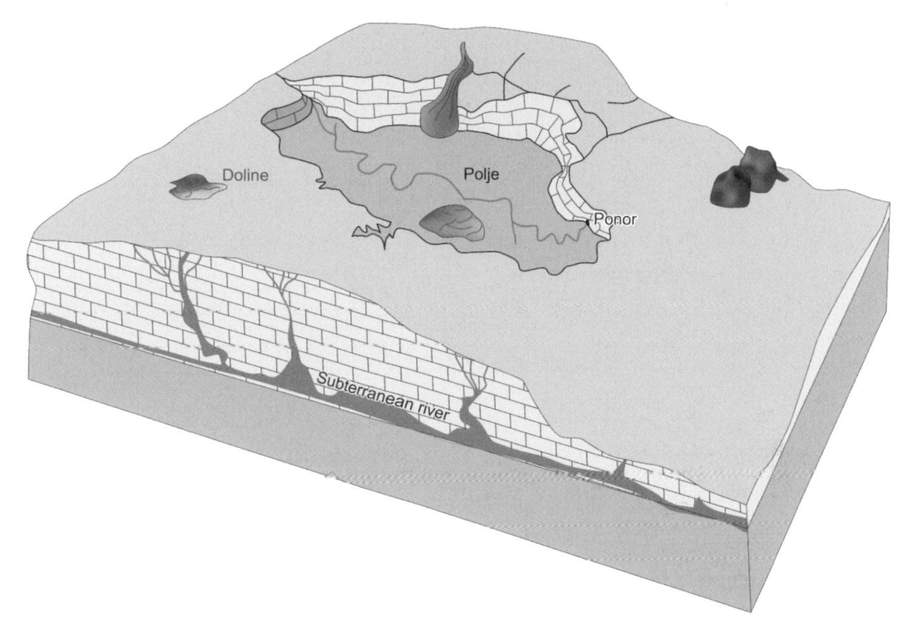

Karst landform created by the dissolution of limestone. The polje was originally a cave; after the roof collapsed a closed depression was left.

Pheneos polje, Achaia, Greece. The ponor (not in the picture) is to the extreme right.

referred to by the term 'karst', named after the Kras or Karst Plateau in Slovenia, on the northeast extremity of the same mountain range whose most southerly outcrops we now see in Greece.

The classical Greek writers understood this all perfectly. That makes the puzzle even more difficult: how can it be that, in a country with so many caves and subterranean rivers, the Greeks located the Acheron and the Styx, two rivers that do not go underground at all, in their Underworld? Why isn't Charon *here*?

Limestone, limestone, limestone. It's everywhere. Two-thirds of Greece is limestone, all of Jerusalem is limestone, everywhere there are caves, dissolved rock, collapsed rock, subterranean rivers. White limestone coasts are in all the brochures for cheap holidays in the sun in the Mediterranean, as limestone makes the sea blue and that attracts tourists. All limestone – and I don't even like limestone.

My love of geology began in the National Museum of Geology and Mineralogy in Leiden. Next door was the National Museum of Natural History, where my uncle, the aforementioned Lipke Holthuis, worked. He was the cleverest biologist in the whole world because, as I said, he knew all crabs, crayfish, shrimps, prawns and woodlice – and even water fleas – personally by name. When he came to stay with us in our wooden cabin on the Brabantse Wal he taught me the names of the weevils and click beetles that came to visit me in my underground sandpit. And sometimes he would take me with him to his workroom in the museum. It had a classical entrance. You went in between two elephant skeleton; halfway up the stairs you found a Pithecanthropus staring at you, which Uncle Lipke irreverently called 'Piet', and the endless granite corridors with their display cases full of dead life smelled vaguely of formaldehyde, alcohol and science. All the walls of his workroom were full of bookcases, the books all standing neatly in line and protected against the dust by green felt pelmets hanging from the shelf above. Darwin was there, just two books, about barnacles. 'His best works', said my uncle. He didn't like evolution: species should not change from one to another, as that would mess up their names. The limitations of an extreme taxonomist who was not interested in evolution or ecology. But I didn't understand that until much later.

Whenever I had the opportunity, I would slip next door to look at the minerals, my real passion. I could spend hours staring at the wonderful geometric forms created by nature itself: cubes, prisms, tetrahedrons, octahedrons, rhombohedrons, beautifully coloured and full of rare elements that nurtured my embryonic interest in chemistry. That

Sulfur crystals, École des Mines, Paris.

The only kind of limestone I like: a calcite crystal ('Iceland spar') showing birefringence, or double refraction; École des Mines, Paris.

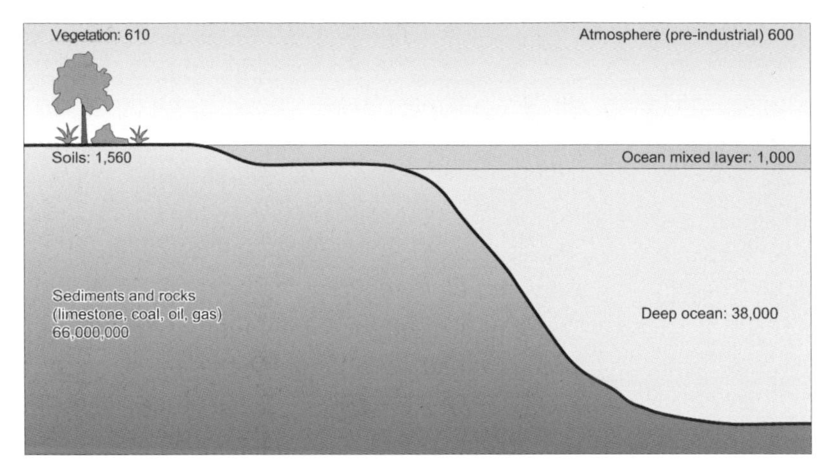

The Earth's carbon reservoirs, figures in gigatons (Gt; 1 Gt = 10^{15} g).

passion has never left me and although I have considerably extended my area of work in geology, I still have a secret love of rare minerals and crystalline rocks.

Limestone consists of only one mineral, calcite, calcium carbonate, and sometimes a little aragonite, a different form of the same compound. Although calcite does have many different crystal forms, in chemical terms it is dull. Half of the world's oilfields are found in limestone, and my students at Delft were keen to know more about it, but I always managed to give the subject a wide berth. In theory I could have made a neat link between my geology and Uncle Lipke's biology. But, it has to be said, because of his aversion to evolution, he showed little interest in fossils. In any case, things turned out differently.

Only the geologist Peter Westbroek succeeded in tempering my prejudice against limestone somewhat with his excellent book *Life as a Geological Force*. Limestone is very special, he says, because it is a product of life itself. And, of course, he is right. The primitive atmosphere may have comprised 90 per cent carbon dioxide. But, in geological history, life has extracted nearly all the CO_2 from the air and stored it in limestone, so that only 0.025 per cent remains. There is 40,000 times more CO_2 in limestone than in all the oceans, the atmosphere and the living biosphere together. If you see what clever strategies life has devised to use that calcium and what fantastically diverse gossamer-thin calcium skeletons drift around in the oceans, you are forced to respect those organisms.

As we saw above, when limestone dissolves in karst areas, both the calcium and the carbonate are transported by rivers to the sea in the form of bicarbonate ions. Bicarbonates are also transported to the sea by the erosion of silicate rocks, such as granite and basalt, by acidified rainwater. That is, however, relatively much less than from limestone as these rocks contain little or no carbonate, and the silicates in the rocks, such as feldspar and mica, do not dissolve but are transformed into other minerals, especially clay minerals. Quartz hardly dissolves at all.

Once the calcium and the bicarbonate arrive in the sea, they discover that it is already saturated with calcium carbonate. Shells and other organisms make use of that to protect themselves against predators. To make calcite again, one calcium ion needs two bicarbonate ions. That is because the H in HCO_3^- has to become water again, H_2O, which requires two hydrogen ions, and therefore two bicarbonate ions, since each bicarbonate ion contains only one hydrogen ion. But calcite only needs one carbonate ion. The CO_2 from one bicarbonate ends up in protective shells, but the CO_2 from the other one disappears back into the atmosphere. The equilibrium in the equation $CaCO_3 + H_2O + CO_2 = Ca^{2+} + 2HCO_3^-$ then moves back to the left. The formation of limestone in seawater therefore releases carbon dioxide! That seems to go against all logic, but it is true. And here, too, subtle variations in the oxygen isotopes of the calcium skeletons are

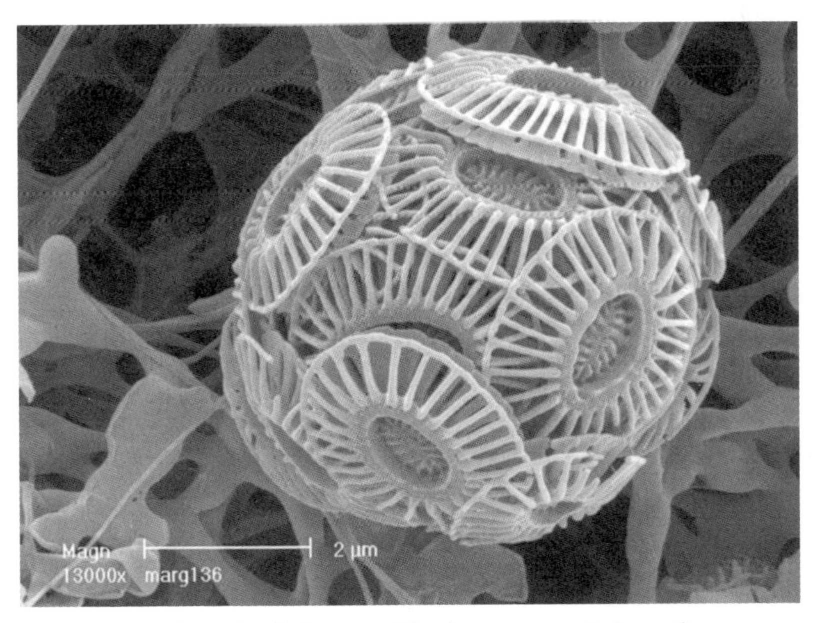

Emiliania huxleyi, a coccolithophore: an organic form of calcium carbonate and, yes, quite beautiful.

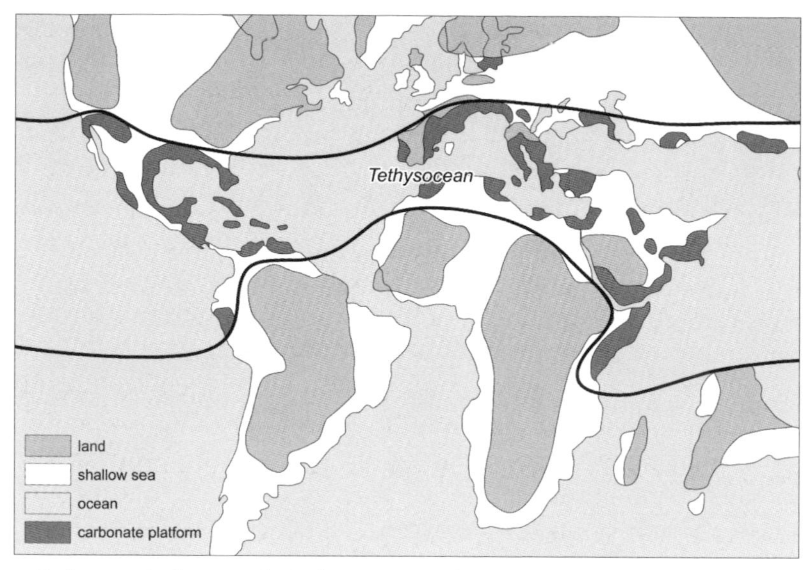

Carbonate platforms in the Tethys Ocean in the Jurassic and Cretaceous periods.

an indication of climate change: they are other chapters in the same history book that can be read in the stalagmites in the dripstone caves.

It is no coincidence that there is so much limestone in the Mediterranean Sea, around which the classical world developed. It is related to the climate and the geological history of plate tectonics in the region. Limestone is not formed at random. By far the greatest quantity of limestone is created in tropical climates, in the form of coral reefs in clear, warm, oxygen-rich water containing no clay sediment. Mud coasts with pure carbonate sediments occur when certain organisms deposit calcium carbonate from the sea water in their slime or through the erosion of old reefs. At the moment, this is happening only in a few places, as the world's climate is too cold: we are in an era when ice ages have the upper hand. We see carbonate mud coasts in the Persian Gulf and the Bahamas and corals in the tropical oceans as long as there are no rivers in the vicinity to deposit their sediment in the sea.

But between 66 and 250 million years ago, the Mesozoic era, the climate was tropical to subtropical practically everywhere, from the equator to the poles. It was especially warm in the Jurassic and Cretaceous periods, and the CO_2 content in the atmosphere was much higher, perhaps five to twenty times greater than now. Moreover, the sea level was 200 metres higher than it is now, because there were no ice caps at the poles. Sixty per cent of the current surface of the Earth lay under water. The Earth was a greenhouse.

In that time the Oceanus of the classics was a real ocean: the Tethys Ocean. Europe and Africa were drifting apart, just as Europe and America are now doing. It was a small piece of the puzzle in Alfred Wegener's scenario of the disintegrating supercontinent Pangea. On the floor of the Tethys Ocean was a spreading ridge, just as there is now on the floor of the Atlantic Ocean, on which Iceland and the Azores lie. In the Mesozoic greenhouse climate gigantic quantities of limestone were deposited on the shallow coasts of the Tethys.

In geological terms this spreading phase did not last long: around 100 million years. It started early in the Jurassic period and ended late in the Cretaceous era, or shortly after. Europe and Africa then started to move together again and eventually collided. If you want to see what the seabed looked like then, you have to go to the Troodos mountains in Cyprus, where it was pushed up above the water level by the collision. The Alps, the Apennines, the Greek Mountains, the Carpathians and the mountains of Turkey, the Caucasus, Iran, Afghanistan and even the Himalayas are the result of the closing of the Tethys Ocean. The collision is not yet complete: the Mediterranean Sea is all that remains of the Tethys Ocean and, in another 50 million years, Europe and Africa will be firmly welded together. Then the Mediterranean, the Black Sea and the Caspian Sea will all disappear, and all that will remain of them will be a deep suture in the mountains of the Cordillera Mediterránea in the supercontinent of Afreurasia.

When the Tethys Ocean started to close, the limestone deposits came under pressure, folded, shifted over each other and rose above the water level, so that we can now see them on the land. Thanks to the folds we can see, in the cores of the anticlines, deeper into the Earth than if Oceanus had opened up further. And thanks to the continuing uplift, we can now find multi-storey karst caves created by underground rivers all over the Mediterranean region.

Who knows: perhaps the Underworld would have looked very different if classical culture had developed around the North Sea.

A late 11th-century portrayal of the Last Judgment on a wooden panel from the Bjarnastaðahlíð farmhouse in northern Iceland, which probably came originally from the cathedral at Hólar. Now in the National Museum of Iceland.

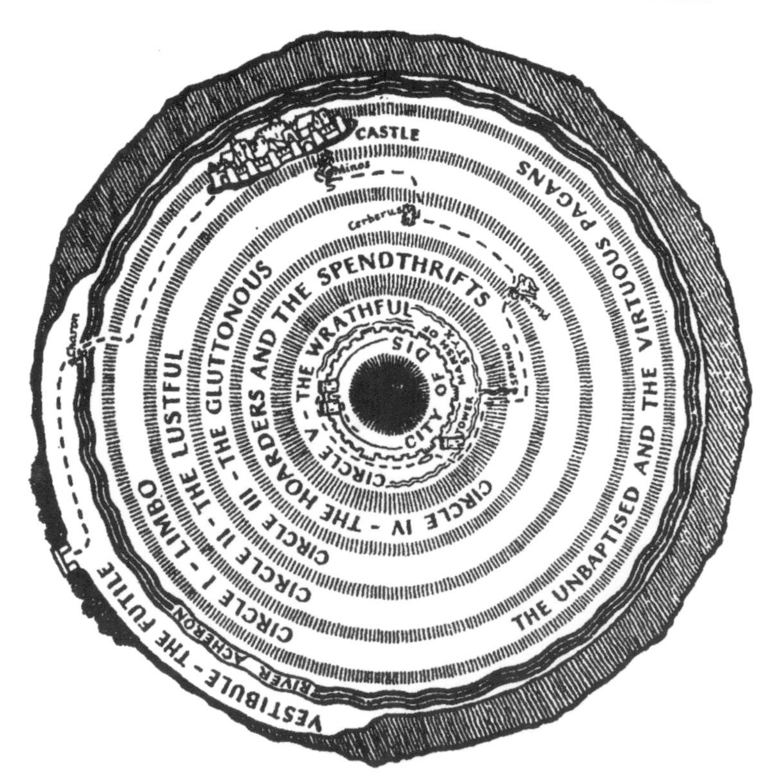

Upper Hell: Dante's Inferno as a gobstopper.

SEVEN

Limbo

'They have not sinned. But their great worth alone
was not enough, for they did not know Baptism
which is the gateway to the faith you follow,

and if they came before the birth of Christ,
they did not worship God the way one should;
I myself am a member of this group.

For this defect, and for no other guilt,
we here are lost. In this alone we suffer:
cut off from hope, we live on in desire.'

The words I heard weighed heavy on my heart;
to think that souls as virtuous as these
were suspended in that limbo, and forever!

Dante Alighieri, *Inferno*, IV:33–45

We haven't got very far with the structure of the gobstopper. You can enter the Underworld at Lago Averno, and the Strait of Kerch, at the Acheron, or at the Necromanteion, and you can be ferried across by Charon, but what then? Homer tells us that under the kingdom of Hades lies Tartarus, but a two-layered globe is a somewhat primitive portrayal of the world beneath our feet. The soil in which our graves lie has more layers than that, and it is only 120 centimetres deep.

The early Christians were inspired by the final judgement in the Revelation of St John, and envisaged the Underworld as full of skulls, worms, fire and devoured sinners, but with no clearly defined geography. They were more concerned with whether hell would continue to exist after the Apocalypse than with what it looked like. Anyone wishing to know more about our views of hell through the ages should read Alice Turner's excellent book, *The History of Hell.*

And then, suddenly, 2,000 years after Homer, Dante comes up with a crystal-clear portrayal of the Inferno in nine concentric circles, divided by subterranean rivers, ravines, walls, steps, dykes and canals: a wondrous reconstruction of the subterranean natural world using the modern architecture of the Renaissance. Each circle has its own sinners and punishments. The first four circles are largely inhabited by those guilty of sins of wantonness: the lustful, the gluttonous and the greedy. In the fifth circle the Styx flows around the city of Dis where, as in the sixth circle, heretics burn in open tombs. The seventh circle, where the Phlegethon – the river of blood – flows, is divided into three levels to separate those guilty of different kinds of violence. The eighth circle, the Malebolge, is the most complex, consisting of ten sublevels known as *bolgie* (ditches), in which those guilty of all kinds of fraud are subjected to the most horrendous punishments. In the ninth circle traitors are frozen in a lake of ice. Right at the bottom, Lucifer resides in the ice of the Cocytus.

It is a system constructed by a man who loved numbers. Dante's *Inferno*, *Purgatory* and *Paradise* form a trilogy, each book containing 33 cantos. If you count the introduction to the *Inferno*, there are exactly 100. He uses a strict metre of three lines, the terzarima or *terzina*, which follows a strict three-line rhyming pattern: a-b-a, b-c-b, c-d-c, and so on.

We've already had occasion to refer to Dante, at the gates of hell, and when crossing the Acheron with Charon. But there is so much detail, so much worth seeing, so much geology, landscape, desert, water, pitch, fire and ice in his *Inferno* that we will follow him as our journey proceeds in this book. You could not wish for a better guide to the Underworld and, although it was written 700 years ago, there is nothing since that can hold a candle to the *Inferno*. But it is advisable to take a map with us.

It would be wonderful if Dante's Inferno had been a gobstopper, but it isn't: it is more of a funnel-shaped pit, in which each circle is smaller than the one above it, and which ends in a point at the centre of the Earth, the seat of Lucifer. It is in effect a kind of negative of the Tower of Babel.

> And then I saw what I had not before:
> the spiral path of our descent to torment
> closing in on us, it seemed, from every side. (XVII:124–6)

There are countless depictions of this layered funnel. The best is by Botticelli, who illustrated the first printed edition of *Inferno* in 1481 with 92 drawings for the Medicis in Florence. As the picture on page 113 shows, modern mine builders seem to have imitated Dante.

This is the best map of Dante's Underworld, and we shall encounter many of these details in our journey. But before we venture into the depths, it is fair to ask the question: where did Dante find his inspiration? As far as the philosophical and literary aspects are concerned, that is no secret. Dante knew his classics: Aristotle, Boethius, Cicero, Virgil of course, Ovid, Horace, Lucanus, Statius and Livy, Biblical and pagan antiquity. Yet none of these writers have anything to say about the highly original structure of the Underworld. The world views of Aristotle and Plato are important in Dante's thinking, but there are no indications that they had any influence on the fantastic cross-section of the gobstopper that Dante presents. Did he make it all up himself?

Until the early twentieth century, that was exactly what people believed. In 1919, however, a book was published in Madrid that was to set the whole Christian world on its head: *La Escatología Musulmana en la Divina Comedia* by the Spanish priest Miguel Asín Palacios. In the book Asín Palacios showed that the structure of Dante's whole *Divine Comedy* was taken from Islamic sources, especially ninth-century Arab legends. Dante! The leading icon of Christianity, the main portrayer of the glories of Paradise and the horrors awaiting us in hell it we fail to lead pious lives! From Islam? Surely that's impossible!

Asín Palacios was prepared for the backlash, as we can see from his preface: 'The above is, in outline my thesis. It will sound to many like artistic sacrilege, or it may call an ironic smile to the lips of those – and they are not a few – who still conceive an artist's inspiration as something preternatural, owing nothing to any suggestion outside itself.'

The story that Asín Palacios refers to consists of two parts, the *Isra* and the *Mi'raj*. In the *Isra*, Mohammed travels in one night to the Furthest Mosque (Jerusalem). From there he climbs a glittering staircase of emeralds, rubies, pearls and other precious stones to the heavens: that is the *Mi'raj* (ladder), Mohammed's ascent to heaven. He is carried on a *buraq*, a winged horse, and accompanied by the angel Gabriel. The distance between each heaven and the next is 500-years' walk. The first heaven is made of iron and, when they arrive there, Gabriel calls for the gate to be opened. The gate-keeper asks: 'Who is there?', to which the angel replies 'It is I, Gabriel.' 'Who is with you?' 'Mohammed', says Gabriel. 'Was Mohammed summoned?' Gabriel replies that he was.

Sandro Botticelli, illustration to Dante's *Inferno*, 1480–95, drawing on parchment: increasingly narrow spirals.

They are welcomed to the first heaven and meet Yza-ibn-Marien (Jesus) and Yohanna-ibn-Zacharia (John the Baptist). At the gate to the second heaven, which is made of copper, the ritual is repeated and they meet Joseph; at the third, made of silver, Enoch and Elijah; at the fourth, of gold, Aaron; at the fifth, of pearls, Moses; at the sixth, of emerald, Abraham; and at the seventh, of ruby, they meet Adam. In the eighth heaven, made of topaz, Mohammed hears the voice of Allah from behind a curtain, asking him to fast for 60 days a year and pray 50 times a day. Mohammed considers that a little excessive and, negotiating smartly, manages to get it reduced to 30 days of fasting and praying five times a day. Gabriel then shows Mohammed around Paradise.

Afterwards Gabriel tells Mohammed that Allah loves him so dearly that he wishes him to see hell and describe to his people what he has seen. The angel says that below the Earth is a land of fire, with a sea of fire and fish of fire. Nearby is another land of fire, with people of fire and fish of fire. All in all there are seven of these lands. In the first the wind blows so hard that it can destroy the entire world, and it will do just that on the Day of Judgment. The wind will skin all sinners and they will be consumed by fire.

Aitik copper mine, Sweden.

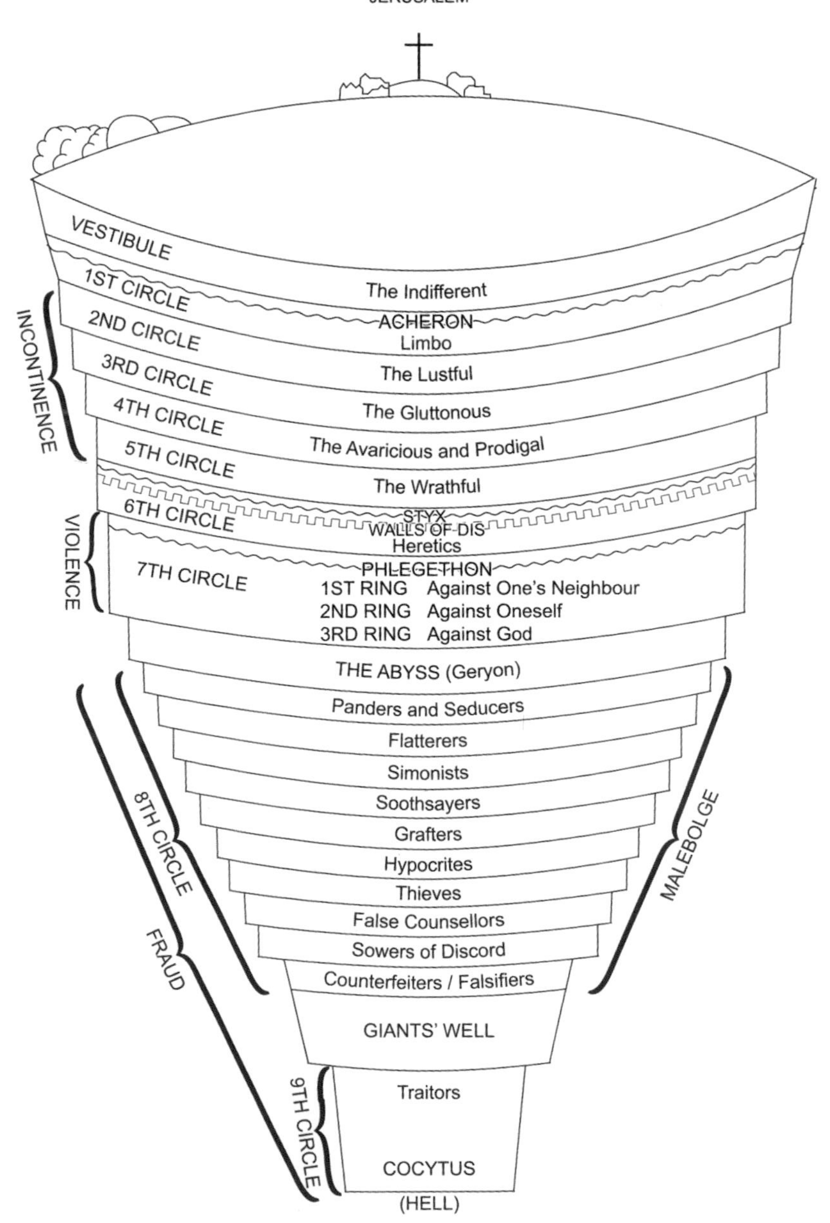

The structure of Dante's Inferno.

In the second land, Halgelada, there are scorpions as large as mules, whose poison can destroy the whole world. The scorpions grasp sinners by their hair, skin them and use their poison to separate the flesh, bones and nerves. In the third country, Area, live demonic beasts as large as mountains, made of earth and fire and blacker than the blackest night, atheir poison is so powerful that it burns more cruelly than the great fires of hell. These beasts grab the sinners, crush them and wrap them in poison from their open mouths, so that they completely melt.

The fourth land, Alhurba, is inhabited by black serpents with 18,000 fangs, each of which is as large as the biggest palm trees to be found on Earth. And each fang has such powerful poison that it can even stir up the fires of hell. The poison from a single fang can destroy the whole world. And, as the sinners discover, when the serpents bite them and release a little of their poison, their bodies disintegrate and their bones and joints break, from their heads to their toenails.

The fifth land, Malca, is full of sulfur stones from hell. 'These are the stones of which Allah speaks in the Koran when he says: "We shall prepare stones to burn and torture the sinners."' The angels of hell tie one of these stones to the neck of a sinner, and take him to the fire. When they get there, the sulfur stones ignite immediately, generating a flame of unimaginable height, and covering the faces of the sinners with fire.

In the sixth land, Zahikika, Allah keeps books and letters recording all the sins people have committed until the Day of Judgment. All their names are listed, what they have done, how, where and why. In this land, too, there are seas of the bitterest waters, known in Arabic as Halmochale, which means abhorrent. Even the beasts of hell are averse to these waters. The water is so bitter that if you dip a rock as large as a mountain in it, it will immediately disintegrate until there is nothing left. Sinners bathe in the water and, if they drink from it, they immediately decompose and their bodies are destroyed.

The seventh land, Hagib, is the kingdom of the Devil, with his people and his armies. He is confined by iron chains, one hand and one foot in front of the other. But he is so big that his head touches the Earth on which we live and his two horns make holes in the ground. The time will come when he will be released and sent into the world.

Gabriel takes Mohammed to a place where he can see hell for himself, with its seven gates, one above the other and all glowing hot. The distance between each of these gates is 70,000 years' walk. Before each gate there is an enormous multitude of demons and people who are in excruciating pain. Mohammed asks how Allah divides the sinners up between the different gates. Gabriel explains that through the first gate, Gehenna, pass those who have worshipped wooden or metal idols, and through the second, Lada, those who have lost their faith in the true teachings of Allah. The third gate, Halhatina, is for those who have amassed wealth unjustly, and for those identified specifically by Gog and Magog. The fourth, Halzahir, is for those who play dice and other games and curse Allah if they lose. The fifth gate, Zakahar, is for those who do not say the right prayers and do not give alms to the poor. The sixth gate, Halgahym, is destined for those who do not believe in the angels and the messengers of Allah, who condemn them as liars and even contradict what they say. The seventh gate, Halkehuya, is for those who deceive and who cheat with weights and measures.

According to Asín Palacios, the similarities with the *Divine Comedy* are striking: in both books, hell consists of circular levels, and the deeper you go, the more severe are the sins the inhabitants have committed. Deceivers populate the deepest levels and the monsters resemble each other. In both the descent to hell takes a day. The guardians of hell are the equivalents of Charon, and many more details are conspicuously similar to those in Dante.

The incredulity with which the revelations of Asín Palacios were received was not only based on indignation about him casting doubts on Dante's originality and his alleged defilement of Christian values, but also because he could not prove how Dante could have known about the *Mi'raj*. In 1949, however, five years after his death, two manuscripts were discovered at practically the same time, one in Latin in the Vatican and the other in French in Oxford, of a book called *Liber Scalae Mahometi*, the most beautiful and complete *Isra* and *Mi'raj* to have been found up to that date. The foreword to both manuscripts states that the book was first translated from Arabic to Spanish by order of Alfonso x by the Jewish doctor Abraham Alfaquim, and that the king's secretary, Bonaventura da Siena, had produced the Latin and French translations. The discoverer of the Latin manuscript, Enrico Cerulli, was able to prove that it had already been in Italy before 1247, and that it was therefore very possible that Dante knew of it when he wrote the *Divine Comedy* in 1300. The Arabic and Spanish versions have unfortunately been lost. The twelfth and thirteenth centuries were a period of intense interaction between Arab and Western culture and science.

My partial summary above is based on the Italian translation of the Latin text *Il libro della Scala* by Roberto Rossi Testa, with a very comprehensive commentary by Carlo Saccone, which was my main source for the history of the manuscript and how it was received. Saccone believes that the history of the *Mi'raj* extends further back than the ninth century: there are allusions to the story in the Koran, which also speaks of the seven gates of hell, and there may be even deeper roots in Iranian and Indian mythology. Dante's wonderful gobstopper therefore seems to have come originally from the east.

But Dante is far from grateful to Mohammed: his description of how the prophet is punished for dividing humanity by announcing a new religion is one of the most gruesome passages in the *Inferno*:

> No wine cask with its stave or cant-bar sprung
> was ever split the way I saw someone
> ripped open from his chin to where we fart
>
> Between his legs his guts spilled out, with the heart
> and other vital parts, and the dirty sack
> that turns to shit whatever the mouth gulps down. (XXVIII:22–7)

Dante is much more friendly to Arab scholars: he had a deep admiration for science. Once Charon has taken you to the other side, you enter the first circle, limbo. That is inhabited by a rather elite company of the dead who may not have committed any sins but who have also not been baptised. They were unfortunate enough to have lived before the birth of Christ, or were born into the wrong faith. A number of leading figures from the Old Testament, like Abel, Noah, Moses, King David, Abraham, Jacob, Rachel and many others, are no longer here: Christ took them with him to Paradise after his descent into hell, as described in the Gospel of Nicodemus. Those who are here include the great poets of classical antiquity, like Homer, Horace, Ovid and Lucanus; the great philosophers, including Aristotle, Plato, Socrates, Democritus, Thales, Heraclitus and Empedocles; and other scholars like Ptolemy, Euclides, Hippocrates and Galen, and the medieval Islamic scholars Avicenna and Averroes. Avicenna (981–1037), who was of Persian origin, was a legendary doctor, philosopher, poet, astronomer and mathematician from the heyday of Islamic science. His *Canon of Medicine* was a standard work for many centuries. Averroes (1126–1198) was an influential philosopher and astronomer in Al-Andalus, Moorish

Iberia. The fact that Dante was well informed about Islamic science makes it more likely that he also knew the *Mi'raj*.

In another, uncompleted scientific text, *Convivio* (The Banquet), Dante quotes the Persian-Arab astronomer Alfraganus, who said that the Earth was 6,500 miles in diameter. He quotes the same number elsewhere in the text and in his *Paradise*. It is not clear whether he meant the Roman mile of 1,473.38 metres, the Florentine mile of 1,890 metres or the Arabian mile of 1,978 metres, but the fact that he specifies a number at all suggests that he had a precise idea of how large hell was.

Dante was by no means the first to try to establish the scale of the Underworld. The lines of Hesiod, a contemporary of Homer, from around 700 BCE, are well known:

> . . . a brazen anvil falling down from heaven nine nights and days would reach the earth upon the tenth: and again, a brazen anvil falling from earth nine nights and days would reach Tartarus upon the tenth . . .
>
> It is a great gulf, and if once a man were within the gates, he would not reach the floor until a whole year had reached its end. (*Theogony*, 720–25, 740–41)

The textual scholars do not agree whether the first and second passages contradict each other, or whether the deep gorge is an even lower level within Tartarus. Perhaps Hesiod added it later.

The American mathematician Andrew Simoson dedicated a whole book to Hesiod's anvil. For those without a thorough grounding in mathematics and physics it is tough going, but it is also a light-hearted book with references to *Alice in Wonderland*, *Winnie the Pooh*, *The Little Prince* and many other familiar tales. He uses the theory of gravity to calculate where the anvil would have to had started from in space to reach the surface of the Earth in nine days. He concludes that it would be one and half times the distance to the Moon, assuming the anvil fell in a straight line. If the anvil had fallen from the orbit of Saturn, the most distant planet known in classical times, it would have taken nearly 9,000 years, as the gravitational pull of the Earth is very weak so far away. If you take the gravitational pull of the Sun instead of that of the Earth, it would take more than fifteen years, still assuming a straight fall. It is more difficult if the anvil is already moving in the beginning, for example, because it has been catapulted away from its origin, because it would then enter orbit around the Earth rather than falling on its surface.

The anvil's fall from the Earth's surface to the bottom of Tartarus is more complicated, since the resistance of the terrestrial rock has to be taken into account, not to mention the rotation of the Earth and the change in gravity the closer you get to its centre. According to Simoson, if you assume that the Earth's mantle is thicker than its crust, and that gravity increases in the crust but then decreases again as you move past the mantle, nine days is a reasonable estimate. Hats off to Hesiod!

Virgil takes a slightly more modest view: according to him, Tartarus is twice as deep as Mount Olympus is high: 5,834 (twice 2,917) metres, not even half the depth of the Earth's crust, where the continents are at their thinnest. Only under the oceans is the crust a meagre 5–10 kilometres thick, but even then it is more than twice as deep as the deepest subterranean river found to date, at the Voronya Cave in Abkhazia, in the Caucasus, which is almost 2,200 metres deep.

The *Mi'raj* exaggerates to the other extreme. Mohammed asks Gabriel about the scale of the seven worlds of the Underworld. Gabriel tells him that each of these worlds is 1,000 years' walk wide and 1,000 years' walk long, and 500 years' walk deep. At a walking speed of 4 kilometres an hour, a years' walk is 35,040 kilometres, almost the circumference of the Earth. A thousand years' walk is 35 million kilometres, meaning that hell covers a large part of the heavens. If you then remember that, according to the *Mi'raj*, it is 70,000 years' walk between the gate of each land in the Underworld, around 2.5 billion kilometres, hell would be larger than our solar system, extending far beyond Saturn. Not a bad prospect for a geologist: it means we win out on the astronomers after all. Arabian mythology, too, is at odds with the science of its own time.

If you assume that Dante had access to all this knowledge, it can be no coincidence that, in Cantos 29 and 30, he gives the dimensions of two ditches in the Malebolge: 22 miles for the ninth and eleven miles for the tenth, while they are half a mile or more wide. No less a scientist than Galileo reconstructed the dimensions of Dante's Inferno in his youth on the basis of these figures. He was 24 when he was asked in 1587 by the Accademia in Florence to investigate which of the earlier reconstructions of hell was correct: the one by the Florentine Antonio Manetti in 1504 or that of Alessandro Vellutello from Lucca dating from 1544. Some critics suggest that Galileo was biased in favour of his fellow Florentine, but French professor Lucette Degryse believes

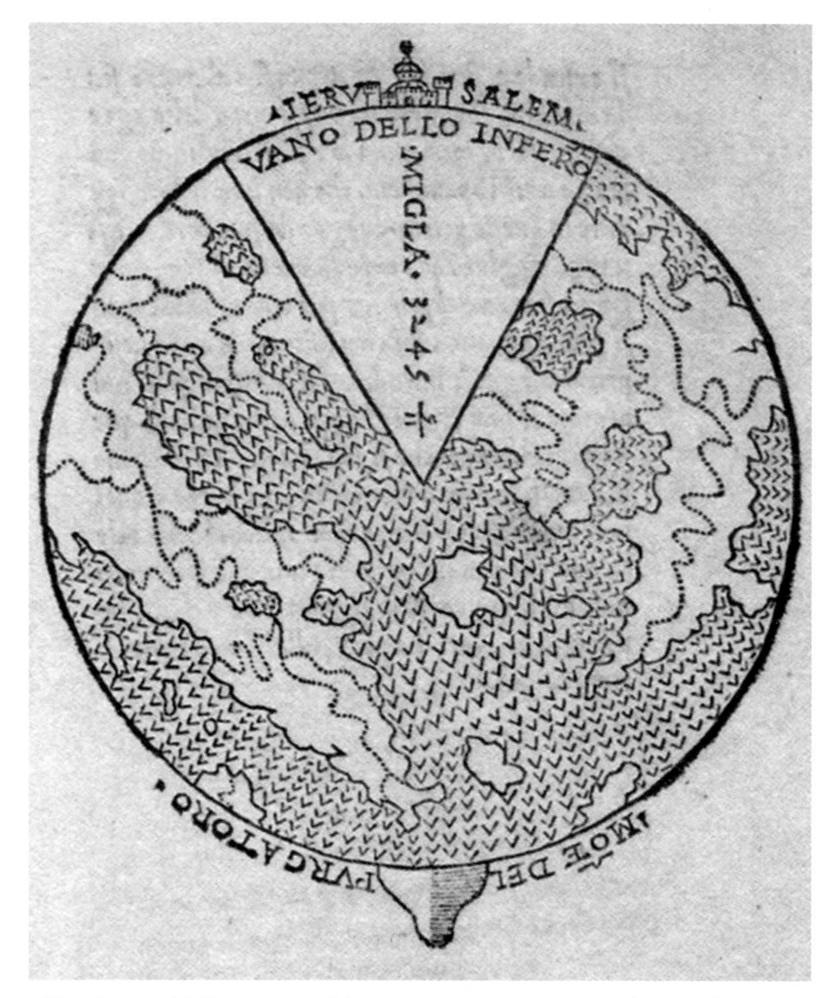

The Cavity of Hell as portrayed by Antonio Manetti in 1504, showing the radius of the Earth (3,245 miles). Jerusalem is at the top and Purgatory at the bottom, in the southern hemisphere.

that they are doing him an injustice. It was an honourable assignment and Galileo hoped that it would give him the credibility he needed to be appointed professor in Bologna. That proved unsuccessful, but two years later he was given a chair in Pisa.

Dante's world view is still based on that of Aristotle and Ptolemy, with the Earth immobile at the centre of the universe and the Moon, Sun, planets and stars revolving around it. Galileo, too, had not yet let go of this traditional view, but it seems reasonable to assume that his study of Dante helped him to radically change that world view later in his life.

Galileo first describes the inverted cone shape of hell: if you draw a circle around Jerusalem with a radius of a twelfth part of the circumference of the Earth, you have the upper side of the cone. Its point is at the centre of the Earth. Using the measurement principles of Archimedes, with which he was familiar, he calculated that the volume of hell is a little less than a fourteenth part of the total volume of the Earth. The cone does not, however, reach the surface of the Earth, because hell has a roof as thick as an eighth part of the Earth's radius.

The vertical distance between the first five circles of the cone of hell is also an eighth part of the Earth's radius of 3,245 miles, Galileo writes, following Manetti. Circles five and six lie at the same level, with no steps between, and the distance between circles six and seven is back to normal. In the eighth circle are the ten ditches of the Malebolge. On the basis of the radii of the ninth and tenth ditches, as stated by Dante, Galileo uses his geometric skills to calculate that the distance between the seventh circle and the first ditch of the eighth circle must be 81 3/22 miles. That is also the distance between the other ditches, and between the lowest ditch and the pit where Lucifer has his seat.

The fact that hell has a roof is implied by the darkness that reigns there. But that is difficult to reconcile with the origins of hell. Lucifer was an angel of God who was expelled from heaven and fell to the deepest point of the universe as it was portrayed at the time: the centre of the Earth. His fall created the hole that became hell. That means that hell is, to a certain extent, an impact crater. The material displaced by the impact now forms a protuberance on the southern side of the globe: Purgatory. But that makes it difficult to understand how hell came to have a roof. Impact craters do not have roofs.

French professor Jean-Marc Lévy-Leblond has checked Galileo's calculations. The volume that Galileo arrives at following Manetti is correct, he writes. The problem is that the roof is by no means strong enough not to collapse under its own weight. There goes another neat symbiosis of myth and science. Perhaps Galileo should have devoted more attention to geology than to astronomy; he was doing so well up to that point.

It is remarkable that the sages assembled in the limbo of Dante's Inferno spent so little of their lives thinking about the interior of the Earth, perhaps even less than the poets who share their fate. In a sense the philosophers were natural scientists: they tried to understand the world on the basis of observations, without assuming any kind of divine intervention. They looked at the celestial bodies and their constantly changing positions in the firmament and developed theories on the structure of the cosmos. Aristotle grouped the elements of Empedocles – water, air and fire – around the immobile Earth, with a fifth element, which he called ether, around them. Above that, the sky was constructed of 55 crystal spheres, on which the Sun, the Moon, the planets and the galaxies were fixed. That view of the universe would dominate for 2,000 years, until Copernicus, Galileo and Kepler showed that it was wrong.

In Miletus, in Asia Minor (where Turkey is today), a school of philosophy developed as long ago as the sixth century BCE which thought about the form of the Earth. We do not know that directly from their texts, but from Aristotle's *On the Heavens*. According to Thales of Miletus, one of the residents of Dante's limbo, the Earth floated on water. It is often claimed that he saw the Earth as a flat disc, but there is no reference to that in the classics and some textual scholars believe that it can be deduced from Aristotle's words that Thales clearly thought that the Earth was a sphere. But his pupils in Miletus had their own ideas. Anaximander claimed the Earth is a short pillar immobile in space, a cylinder three times wider than it is long. The inhabited world is on the top of the cylinder, and is surrounded by Oceanus. Anaximenes, a pupil of Anaximander, said that the Earth had the shape of a disc, and Leucippus a drum. In the fifth century BCE,

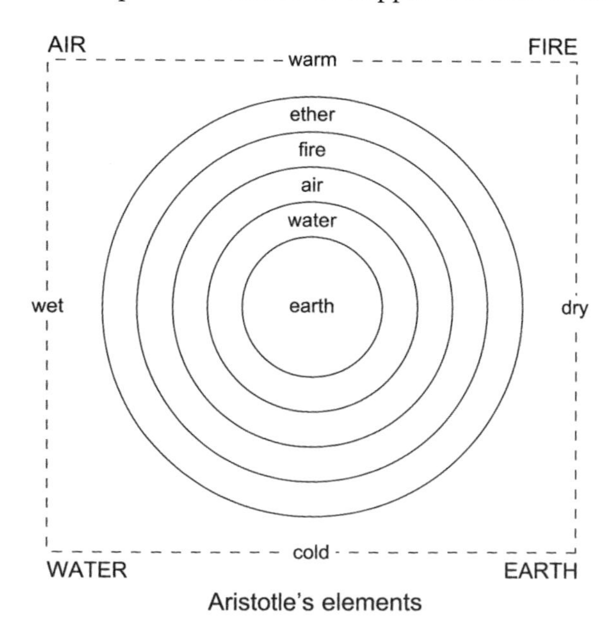

Aristotle's world view.

the Thracian Democritus, who named the atom and is also a resident of Dante's limbo, described the Earth as a hollow ring, a little like a car tyre.

In the third century BCE, however, Eratosthenes showed that the Earth is a globe. For Aristotle, this was self-evident. He writes something much more interesting about the Earth: 'If the Earth was generated, it must have been formed in this way, and so clearly *its generation was spherical*' (my italics). Its generation was spherical! The beginnings of the gobstopper! I would love to press Aristotle further to find out exactly what he meant by that.

Plato, Aristotle's mentor, has Socrates say in the *Phaedo*:

> All have numerous perforations, and there are passages broad and narrow in the interior of the earth, connecting them with one another; and there flows out of and into them, as into basins, a vast tide of water, and huge subterranean streams of perennial rivers, and springs hot and cold, and a great fire, and great rivers of fire, and streams of liquid mud, thin or thick (like the rivers of mud in Sicily, and the lava streams which follow them), and the regions about which they happen to flow are filled up with them. And there is a swinging or see-saw in the interior of the earth which moves all this up and down, and is due to the following cause: there is a chasm which is the vastest of them all, and pierces right through the whole earth . . . And the see-saw is caused by the streams flowing into and out of this chasm. (111d–112a)

The largest chasm in the interior of the Earth is the Tartarus. Plato compares the flowing in and out of the rivers with respiration. Water seems to be the main feature of the interior of the Earth and fire only secondary, and it remained that way for almost 2,000 years. Quite logical, you might think, since water springs from the ground everywhere, while volcanoes only occur here and there.

Plato's view that the interior of the Earth was filled with water held sway for nearly two millennia. Leonardo da Vinci was a fervent supporter of his ideas and provided new evidence of his own:

> It is assumed that all elements must stand in ten-fold relation to each other and that the height of the mountains must be as great as the depth of the oceans. If one were to fill the sea

> with the earth of the mountains that rise above the sea, then the earth would be completely covered by the sphere of water. This water, which would completely cover the entire sphere of the earth, would not reach the summit of the highest mountain. Under these assumptions, this means that water must still be located under the surface, namely in the lower-lying underground streams.

All his writings on water are saturated with Plato's metaphor of the respiration of the water cycle.

Leonardo was also the first to draw the gobstopper, in his Codex Leicester. It looks more like a peanut brittle or a scrunched-up ball of paper than a gobstopper and, if he had not explained that it was a cross-section of the Earth, no one would have been any the wiser.

It is a striking contrast to the fantastic paintings of Dante's Inferno from the same period, such as the magnificent fresco by Orcagna in Santa Maria Novella in Florence, which you might say is virtually true to life. Countless depictions of the Apocalypse include references to Dante. One can hardly believe that these portrayals of the Underworld didn't provoke scientists earlier to wonder what the interior of the Earth really looked like. Or perhaps, as Alice Turner writes in *The History of Hell*, Dante's lifelike portrayals made it easier for the intellectuals of the Renaissance and the Enlightenment to reject the existence of hell. It took the independent mind of Leonardo da Vinci to go beyond the *Divine Comedy* in thinking about the interior of the Earth. Leonardo was born too late to be taken into Dante's limbo, but its inhabitants would certainly have listened to him.

Leonardo wrote next to his ball of paper:

> This is meant to represent the Earth cut through in the middle, showing the depths of the sea and of the earth; the waters start from the bottom of the seas, and ramifying through the earth they rise to the summits of the mountains, flowing back by the rivers and returning to the sea. The great elevations of the peaks of the mountains above the sphere of the water may have resulted from this that: a very large portion of the Earth which was filled with water that is to say the vast cavern inside the Earth may have fallen in a vast part of its vault towards the centre of the Earth.

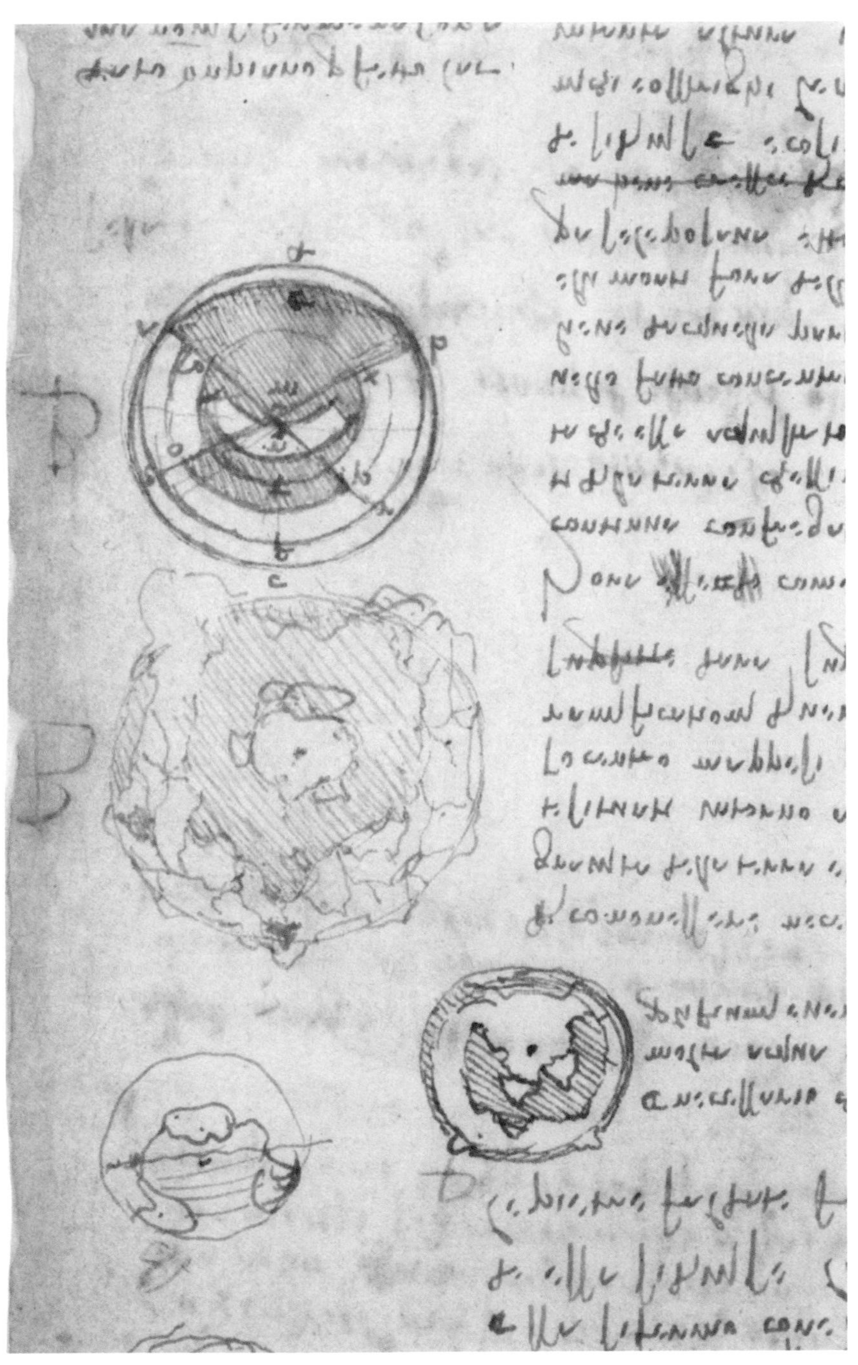

Leonardo da Vinci, drawing from Codex Leicester (1504–10). The vague sketch in the centre is the first gobstopper; the drawing below shows how mountains were formed when part of the Earth's crust collapsed.

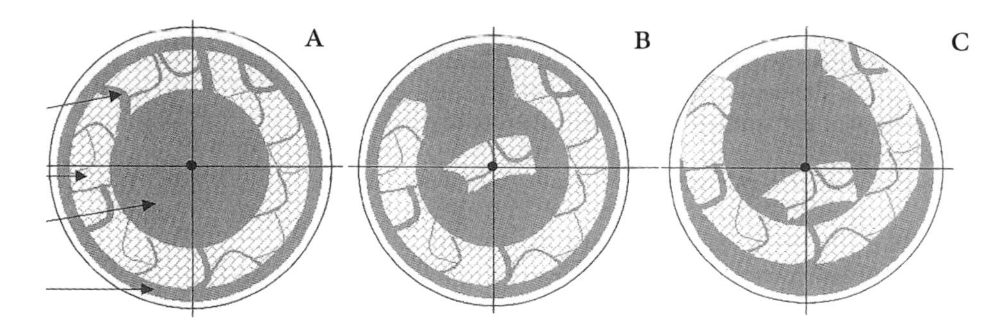

Part of the crust collapses into the underlying layer of water.

If you look closely, you will see that on the underside of the ball there are no mountains. A piece of land has sunk into the depths. Below the ball there is a second drawing that seems to clarify the process.

Hydrologists Laurent Pfister, Huub Savenije and Fabrizio Fenicia recently devoted an entire book to Leonardo da Vinci's theory of water. Their interpretation of Leonardo's drawing is shown here.

The continuous erosion caused by water flowing through the subterranean courses eventually causes part of the roof to collapse and sink down to the centre of the Earth. That causes the surface of the Earth to shift; the pieces that are left become lighter and partly rise above sea level as mountains. The American hydrologist Richard Heggen, who wrote a fantastic book entitled *Underground Rivers*, gives a similar interpretation.

Leonardo's work remained unpublished and thus unnoticed for centuries. It was not until the nineteenth century that it became clear just how much ahead of his time his thinking was, despite the fact that, in this case, he had come to the wrong conclusions.

It is another century and a half before we find another cross-section of the Earth, this time by René Descartes, in the fourth part of his *Principia philosophiae*, published in Amsterdam in 1644. The French translation by Abbé Claude Picot, a friend of Descartes, was published in Paris in 1647 by Henri Le Gras under the title *Principes de la philosophie*. Descartes himself supervised the translation and was probably responsible for many of its deviations from the original, though the addition that God created the world was almost definitely introduced by the abbé. It is a highly theoretical argument owing more to Aristotle than to modern science. André Bridoux, who presented the Pléiade edition of Descartes' main works, did not want to include these 'laborious workings' because 'the Cartesian method did not yet possess the maturity or force to free itself from enormous amount of nonsense under which science was still buried.'

Descartes thought that the Earth was formerly like the Sun. Sunspots were areas where 'the first element' clustered together to form a thicker mass and then thinned out again, so that the sunspot disappeared. He believed that the same process was much more advanced on Earth and had become irreversible. The interior of the Earth was more or less the same as the Sun (I in Descartes' drawings) but, through the clustering of material, a second layer (M) of coarser particles was created, and a third was formed from that (A, B). Through processes that we can best describe as the condensation and separation of particles of different sizes and shapes, further layers (C, D, E and F) are created. These processes are caused by pressure from the light and heat of the sun. It is just like when you press your foot into a marsh and the water separates from the peat, he says in one of his rare arguments based on nature.

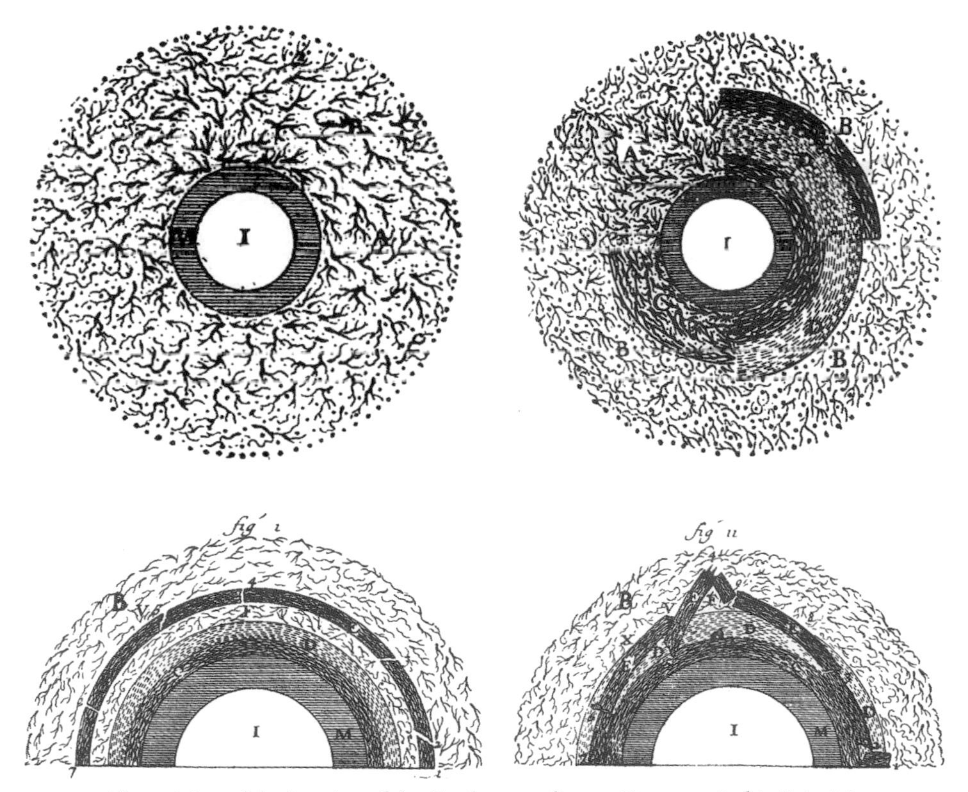

The origins of the interior of the Earth, according to Descartes in his *Principia philosophiae* (1644).

After 30 pages of difficult-to-decipher arguments, all is revealed: B and F are air, D is the water, C is an innermost, very dense crust containing the ores, and E is a another, less solid crust consisting of rocks, clay, sand and loam. Between the outer crust E and the inner crust, C, there is a layer of air, F, and of water, D. The outer crust, E, can develop cracks like those in a dried-out bog. A piece of broken-off crust can sink down through the underlying air and water layers and come to rest on the lower harder crust, C; later fragments will settle unevenly on top of earlier sunken pieces, forming mountains.

This perspective, theoretical as it is, contains much of interest. It is striking that the exterior of the Earth still closely resembles that of Plato and Leonardo. At some places, there is water and air between the Earth's outer crust and the underlying rock, something that it is difficult to see as anything other than a partially collapsed cave. Yet at the centre of the Earth there is something glowing that looks a little like the Sun, and which can separate material: Descartes was the first to assume that the core of the Earth was hot. It would be a long time before there would be more clarity about this.

The collapsing caves continue to be popular. In the sketches that Danish scientist and later bishop Nicola Steno included in his geological account *De solido intra solidum naturaliter contento dissertationis prodromus* (1669), the collapsing process played a prominent role. Steno was the first to formulate the geological law of superposition: that a higher layer is younger than the layer below it – a law I still apply daily to the piles of paper on my desk.

Thomas Burnet (1635–1715), royal chaplain at the court of William III of England, had another explanation for why there must be a layer

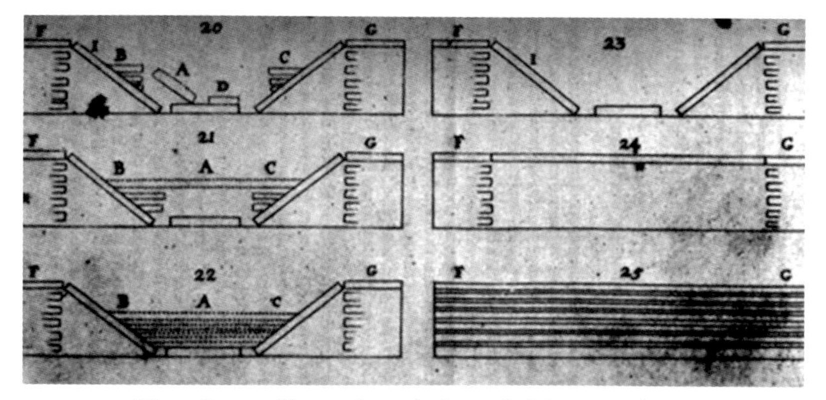

The collapse of layers through the underlying water layer, according to Nicola Steno, 1699.

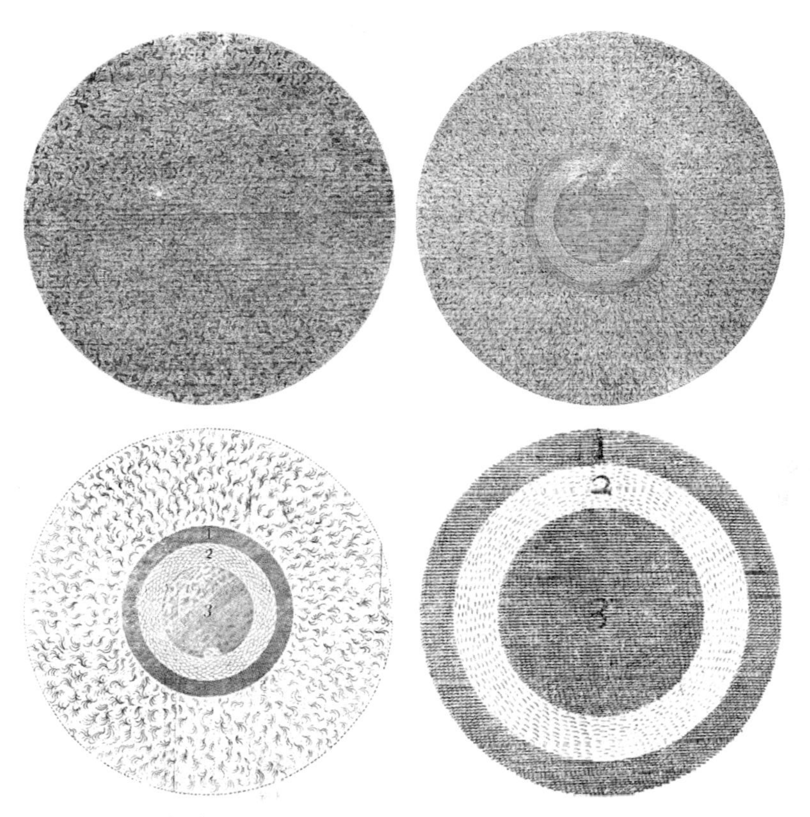

Four stages in the formation of the Earth through the sinking and separation of terrestrial particles (3), oily and watery liquids (2) and air (1), from Thomas Burnet's *Sacred Theory of the Earth* (1681).

of water deep in the Earth. His book *The Sacred Theory of the Earth* from 1681 was long considered an imaginary history of the world, intended to bring it in line with the Bible, but in *Time's Arrow – Time's Cycle* Stephen J. Gould has shown clearly how many of his ideas are actually corroborated by observations from nature. According to Burnet, before the Great Flood, the Earth was Paradise: a smooth sphere with no mountains, no seas and no caves. It had been created by the clustering of particles out of the chaos, the coarsest of which sank first and formed the core of the Earth: this was similar to current theories on the origins of the gobstopper, as we shall see. During the Great Flood, the Earth cracked and large sections sank into the underlying layer of water, producing the chaotic array of mountains, seas and caves that make it what it is today.

How did Burnet know that there was a layer of water under the surface of the earth?

> 'Tis true, all Subterraneous waters do not proceed from this original, for many of them are the effects of Rains and melted Snows sunk into the Earth; but that in digging any where you constantly come to water at length, even in the most solid ground this cannot proceed from these Rains or Snows, but must come from below.

There is groundwater everywhere, so there should be a water layer under the Earth. It is certainly a logical conclusion.

Even at the end of the seventeenth century, Plato's influence was still making itself felt, despite new insights by Newton, Huygens and many other scholars of the Enlightenment. The most splendid gob-stopper from that period is perhaps that of Thomas Robinson, rector of the parish of Ousby, Cumberland, in his book *Anatomy of the Earth* (1694), which not only shows the subterranean water courses, rivers and oceans, but is also reminiscent of the polyhedral shapes of which, according to Plato, the cosmos is formed.

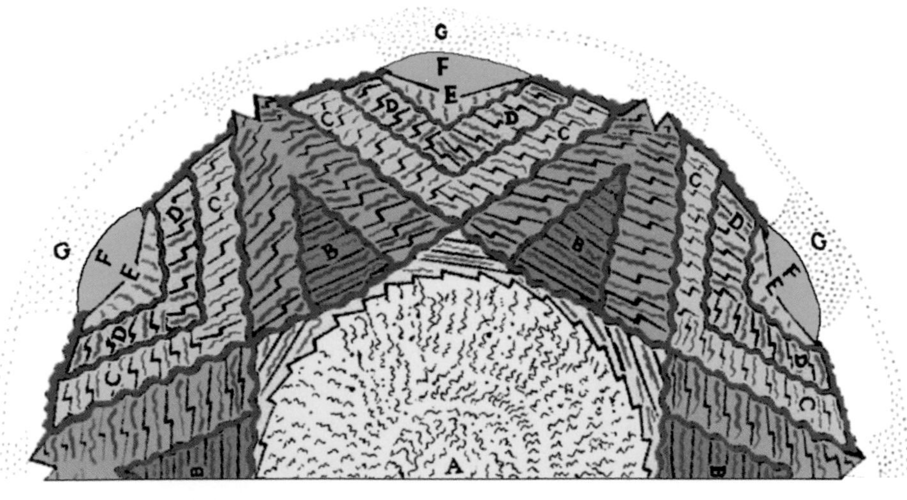

From Thomas Robinson's *The Anatomy of the Earth* (1694) and *New Observations on the Natural History of the World of Matter* (1696); colours added by Richard Heggen. (A) Central fire; (B) Mountains; (C) Heaths; (D) Plains; (E) Channels of the sea; (F) Seas with rivers flowing into them; (G) Vapours arising from the sea. The image shows 'The windings and turnings of the greater Veins . . . through which the whole mass of subterranean Water Circulates. The Lesser Fibres, or Rami Factions, filling all the flat Strata with feeders of Waters, which breaking out upon the Surface of the Earth cause Springs. And thus, in our Bodies, 'tis much easier to break a Vein in the Neck or Arm, where they lye nearest the skin; than in the Buttocks, or any other such Fleshy-part.'

You might wonder where all these collapsing layers of earth – and all that subterranean water and those gradually widening subterranean courses – come from. They are all typical features of karst landscapes in limestone: the springs, the ponors, the cracks in the rocks made ever wider by dissolution, the dolines and the poljes. It is as though, for 2,000 years, the limestone regions of Greece were used as a model for the origins of the Earth.

EIGHT

The City of Dis

For scattered everywhere among the tombs
were flames that kept them glowing far more hot
than any iron an artisan may use!

Each tomb had its lid loose, pushed to one side
and from within came forth such fierce laments
that I was sure inside were tortured souls
Dante Alighieri, *Inferno*, IX:118–23

There is a fascinating paradox in our fear of the Underworld. Caves are not only frightening but have also, since time immemorial, offered us protection from the cold and rain. Our distant ancestors liked to live in caves. People lived in the Vallonnet cave in the Jurassic limestone of the French Alps as long as 900,000 years ago. The Sima de los Huesos, the Pit of Bones, is a dried-out karst river in the Cretaceous limestone of the Sierra de Atapuerca, in northern Spain. Human remains at least 850,000 years old have been found in the pit. Peking Man lived in a karst cave in Ordovician limestone near Zhoukoudian, China, from around 500,000 years ago until the cave collapsed 230,000 years ago. The first Neanderthal was found in the nineteenth century in a karst cave in Devonian limestone near Düsseldorf, Germany. The famous paintings in the karst caves of Lascaux and Chauvet in southwestern France and Altamira, Spain, are 10,000 to 35,000 years old, while the 38,000-year-old Venus of Hohle Fels was found in a cave in Jurassic limestone. The forefathers of the ancient Greeks lived in caves in the Pindus mountains and, according to Anna Petrochilou, 200 families in Greece still lived in caves as recently as 1985.

Not all ancient peoples were fortunate enough to have a karst landscape. Since camping out in the rain was no fun, they built huts. Some, like the early inhabitants of the Netherlands, used turf, others wood or loose stones. If there were no trees, as in the tundra in northern Siberia,

Cave dwellings in Cappadocia.

they fashioned tents from bearskins, with mammoth tusks as poles. The Anasazi, of what is now the southwestern United States, built houses under overhanging cliff faces. Others, however, came up with the idea of making their own caves. Although you cannot hack caves from all sorts of rock, some softer rocks are perfect. Loess and tuff – as we can see in the Neapolitan tuff cave systems around Lago Averno are especially suitable. Both have the advantage that they are not layered and stay dry, because rainwater is captured in the fine pores rather than trickling through cracks and dripping onto the ground.

Perhaps the most famous artificial cave dwellings are in Cappadocia, in the heart of the Anatolian plateau. The area has been a centre of volcanic activity for more than 25 million years and several volcanoes, including Erciyes and Hasan Dağ, are still active. Between 5 and 11 million years ago the region experienced awe-inspiring eruptions, comparable to those in the Campi Flegrei near Naples. At least two old calderas, Derinkuyu and Acigöl, bear witness to this turbulent period. Though these super craters are more than 20 kilometres in diameter, they are now difficult to recognize on the surface. Massive pyroclastic flows covered the entire landscape with thick layers of volcanic glass particles, thc ignimbrites we encountered earlier. Most ignimbrites are white, soft and easy to excavate, though in some places the glass particles have become welded together and become hard. In many places the ignimbrites were later covered by lava flows. Over the centuries, erosion undermined the hard lava on top of the soft ignimbrites, forming a fantastical landscape of white cone-shaped pillars, some still with a cap of dark lava. They are known as earth pyramids or, much more romantically, fairy chimneys (*cheminées des fées* in French). There are whole fields of these pillars, which look like beds of nails for giants. Sometimes they look like melting glaciers, or a procession of portly ladies-in-waiting in hoop skirts, or oversized chess pieces in a perpetual stalemate. Those who prefer to see erotic symbols can let their imaginations run wild. The strange formations are world famous, not only because of their fairy-tale appearance, but also because whole subterranean cities have been created within them, with thousands of houses, churches, monasteries, hermits' cells, dungeons, storerooms and wine cellars. The underground city of Derinkuyu has no less than twenty levels and extends 40 metres below the surface. Some 10,000 people could have lived there. Everything was connected by an incredible labyrinth of narrow passages which, if necessary, could be sealed off by rolling a millstone across the entrance. There was also an ingenious ventilation system, making sure that everyone had fresh air, even at the lowest levels.

What drove people to live underground in this magnificent landscape? Certainly not nostalgia for the limestone caves: the oldest inhabitants of the area, from the Stone Age, did not live below ground at all. This region was part of the green crescent where, 10,000 years ago, when the climate improved after the last ice age, agriculture first developed. Excavations show that people then preferred to live on fluvial terraces, close to the water but on dry land. The Hittites of the Late Bronze Age, the Persians from the sixth century BCE, and later the Greeks and the Romans all lived above ground. People only started to dig below ground in the Byzantine period.

It was essentially a survival strategy. In the sixth century CE the Byzantine Christians found themselves caught between the Arabs advancing from the south, the Persians in the east and, in the ninth century, the Seljuqs (Turks), who conquered the whole land. In the inaccessible landscape of Cappadocia, their subterranean cities protected them for several centuries. It was largely the monks who encouraged them to go underground, which may explain why there are so many churches and monasteries. Or perhaps the inhabitants felt themselves threatened by their physical proximity to the Underworld. Either way, it was not difficult. You could make a cave dwelling in a couple of months with no real knowledge of building. The arches and pillars were based on Byzantine architecture, but were not actually necessary. If you wanted the ceiling round, you made it round. If you wanted it rectangular, you could do that, too. And if you made a mistake, you could always start again somewhere else. The Christians were safe here until the tenth or eleventh centuries. Today no one lives there. The ruins are now used only to store fruit, wine and grain, but you can pay to stay in a cave hotel.

There are, of course, other options, as we have seen in Jerusalem. The city was built using limestone excavated from Solomon's Quarries, under the city itself. And in Naples, with its renowned yellow tuff, the same system has worked for more than 25 centuries.

A summery Sunday morning. The luxury Caffè Gambrinus is already open. Reflected in the gold-framed mirrors, its extravagant sweet pastries entice the hungry passer-by. But that is not why I am here. I am meeting up with a group of fellow tourists curious to see what Naples looks like below the ground. Our guide, Salvatore Quaranta, quickly leads the group – several Italian couples, a Danish family and myself – to the Quartieri Spagnuoli. The Spanish quarters are the most notorious areas of the old city centre, a maze of steep streets paved with *piperno*, the local name for the lava from Vesuvius. There is nothing here to suggest that another city lies beneath our feet.

On a small square, Salvatore suddenly disappears through a door into a house, urging us to follow him. We find ourselves in a small, dark office, with a dilapidated old desk and a wall full of cuttings about the LAES Napoli Sotterranea, the Libera Associazione Escursionisti Sottosuolo, which Salvatore set up with his brother Michele. I don't get much time to study the cuttings, as we are led through a door to a spiral staircase that takes us down 200 steps to 40 metres under the ground. The noise of the endless stream of Vespas and Cinquecentos racing through the city streets is suddenly gone.

We are in a high, vaulted chamber hacked out of the *tufi gialli*. The patterns traced in the tuff are typical of Roman excavation work. And the Romans were not the first to dig out the yellow tuff. In the fifth century BCE the Greeks built the new city of Neapolis, where they found the same yellow tuff as at their earlier settlement of Cumae. They used it to construct their houses and temples, and Naples continued to be

Descent into the negative of Naples through an old mineshaft, adapted in the Second World War to provide access to the underground shelters.

built exclusively from tuff from its own bowels until the Second World War. Salvatore shows us how that was done. Blocks of stone were hacked out of the subterranean rock and hoisted to the surface on a small lift, through a narrow shaft no more than 1 metre across. In this way, a house made a small hole in the ground, and a *palazzo* a larger one. This was intentional, since the hole served as a cistern, a reservoir to store drinking water: a small reservoir for a house and a large one for a whole palazzo. This is known, elegantly, as the 'negative' of Naples. The larger the building, the larger the hole it left in the ground. The water came from small rivers flowing down from the surrounding hills and directed to the ground beneath the city via aqueducts. The cisterns were connected by passages so narrow that heavier-set visitors have difficulty in squeezing through them. The excavations did not go deeper than 40 metres, otherwise they would have reached sea level and salt water would seep into the cisterns.

The area we visit is only a small part of the subterranean city, which is nearly as large as the one above ground. Geologist Mario Tozzi, extensive researcher of everything subterranean, describes his visit to the catacombs of the Fontanelle in his book *Italia segreta*. The negative of Naples is not only a source of building material or a reservoir for drinking water, but also a cemetery, a storage cellar, a cesspool, a sewer, a hiding place for fugitives, a sanatorium for those who believe they can be cured through *speleoterapia*, a church and a place of pilgrimage.

If you died of the plague that had the city in its grip in 1656, Tozzi wrote, you would probably have met your end in the street and been left in one of the tuff pits, since all the churches and crypts were already full of bodies. You would have been placed in a niche in a sitting position to allow all the evil fluids to flow out of your body. Only after you were completely drained would your dried-out corpse be hung in the underground cemetery. When you enter the catacombs of the Fontanelle, the first thing you see in the semi-darkness is incredible rows of skulls resting on a metres-thick wall of shin bones, thigh bones, forearm bones and other long bones. Here lie the remains of at least 40,000 bodies, some claim as many as several million, men, women and children. Researchers have dug 15 metres deeper from this point and found only bones, bones and more bones: the remains of poor people who died from sickness, poverty or an eruption from Vesuvius and could not afford a proper burial. By 1837 the catacombs of Fontanelle were so full that there was no room for any more.

It is now a place of pilgrimage: as well as bones, it is now full of small burial chapels and relic cabinets of marble or wood, containing one or more skulls. Anyone who sees the ghosts of dead loved ones in their dreams adopts a skull in Fontanelle, places letters, flowers, photos or drawings in it and entrusts it with their innermost desires. Tozzi was shameless enough to have read one of these letters: 'Dear soul, come to me in my dreams, tell me your name, and please let me draw the winning lot in the national lottery.'

In the part that Salvatore shows us, there is none of this, only the traces of another horror: the Second World War. Naples was the most bombed city in Italy because of its ports, railways and factories: first by the British, then by the Americans and, after being taken by the Allies, by the Germans. A total of 28,000 bombs fell on the city, causing 20,000 deaths and reducing large parts of the city to rubble.

But thousands of the city's inhabitants survived, fleeing underground by turning one of the narrow lift shafts into a spiral staircase: the same one we use to descend. At least 4,000 people lived there under the ground for months in 500 or more subterranean chambers. Salvatore shows us the rusty remains of water pipes, electricity cables and now defunct insulators. Graffiti scratched on the walls (this is a pleonasm for Italians, as the word graffito means 'scratched') show caricatures of Hitler, Mussolini and Hirohito, aircraft dropping bombs, and, in a niche, the text '*Riservato al signor Campagna*', reserved for Mr Campagna, almost certainly an unpleasant person, and '*Anna e Renzo si sposarono il 20 settembre 1943*', Anna and Renzo married on 20 September 1943, the time of the heaviest bombing of the entire war. The subterranean hiding places were safe; the roof had always withstood the bombs. Hell was up there, in the city.

Much of the subterranean city is now inaccessible because after the war, much of the rubble was disposed of in the underground passages. Only in recent years have pioneers like Salvatore and Michele Quaranta recognized the value of the negative of Naples.

Besides tuff, you can also carve out dwellings in loess. Loess is a fine-grained sediment deposited by the wind, similar to the Sahara dust that you sometimes find in a thin red layer on your car after a strong southwest wind. During our first-year field trip to the Dutch province of South Limburg, our teacher Hans Wensink, a man who was great in the most literal sense of the word, taught us how to distinguish loess in the Edelman auger. 'Loess feels just like the thighs of a young girl', he said with a big grin. I was nineteen and green, and had never felt a young girl's thighs, so I learned it the other way around: that young girls' thighs feel just like loess.

The loess layer in Limburg is only thin, mostly no more than a few tens of centimetres and never more than a couple of metres, but on the Chinese loess plateau in the upper and middle reaches of the Yellow River the loess is a good 300 metres thick. It was blown out of the arid Gobi desert during the ice ages and transported over great distances. Each ice age left its own deposit of loess, covered with a layer of soil dating from the warm period between that ice age and the next. The traces of no fewer than 30 ice ages can be found in the Chinese loess plateau. The plateau is as large as Spain and an estimated 40 million people still live there in caves dug into the loess.

They prefer to dig their caves into steep slopes facing south to get as much sun as possible. Sometimes you see dozens of cave dwel≠lings at the same level in an especially thick layer of loess; all are the same, with round ceilings and flat entrances, like in a dovecote. Often the opening is neatly closed off with a semicircular window and a door. In flat areas with no steep slopes they dig a square hole in the loess ground, like an empty swimming pool, creating a rectangular courtyard. In the four walls of the pit, they dig out separate dwellings that face the courtyard. Whole extended families live in these complexes. If you look out across the plain, you see no signs at all of human habitation, as nothing sticks out above the ground. At the most, there will be a tree that seems to have no trunk, planted in the middle of the sunken courtyard. The subterranean houses are cool in the summer and warm in the winter, and of course, cheap to make. Their only disadvantage is that they collapse during earthquakes, because the cohesion between the loess particles is shaken loose. An earthquake in 1920 caused more than 15,000 cave dwellings to collapse, with the loss of thousands of lives. Yet they were still safer than the above-ground loam houses in the same region, all of which collapsed. The system has existed for at least 4,000 years, and was one of man's first forays into the underground.

But what drove these people to live underground? According to American architect Gideon S. Golany, one of the reasons for the enormous expansion of cave dwellings in the Ming and Qing dynasties from the fourteenth century onward was that in earlier eras so much wood had been used to build houses and for fuel that the area had been completely deforested. Here, too, the choice was made not only for reasons of comfort, but also from bitter necessity.

Nowhere do cave dwellers live closer to hell than in Xinjiang, in the extreme northwest of China. Their ramshackle houses are dug into the thin layer of loess that covers folded layers of grey, yellow and red sandstone and shale. Some have a roof of corrugated sheeting; others are just deep enough not to need extra roof protection. So why have these people chosen to live in this arid, desolate landscape? Agriculture is possible only in the valleys, as the rivers transport smelt water from the snow-covered peaks of the Tien Shan northwards, until they eventually run dry in the vast Dzungarian Desert. Down in the valleys there is vegetation and the villages are of better quality, but here on the mountain sides nothing grows. Those who live here make their living as coalminers.

China lives on coal. The ancient steam locomotives run on coal. In the waiting rooms of stations there are piles of coal. It is everywhere: in car parks, bus stations, courtyards and on the verges of the road. Brickworks, lime kilns and steel factories are fuelled by coal, and the inhabitants of villages and towns burn coal briquettes. The whole country smells of it. In the east there are clean, modern mines but here in Xinjiang they are still extremely primitive.

The layers of coal are wedged in between the sandstone and shale: they are the carbonized remains of gigantic peat bogs that lay here in the Jurassic period. As a result of the folding of the layers, a distant echo of the collision between India and the Eurasian continent, the coal layers now lie either on the surface or deep under younger layers. The coal layers on the surface are sawn into large square sections, prised loose and placed in perpendicular slices on rectangular carts with bicycle wheels. The deeper layers are hacked out in primitive mineshafts and carried to the surface on small trucks running on narrow tracks, sometimes driven by manpower. I saw women pulling these trucks full of coal out of the mines: prostitutes from Shanghai undergoing punishment, I was told. That was in 1995. Perhaps it is better there now.

But that is only part of the problem. If you look out across the coal region from a high vantage point, you can see plumes of smoke everywhere: the coal is burning, sometimes spontaneously, sometimes as a result of the mining. It is one of China's greatest problems. Not only is the country losing large quantities of a very valuable natural resource, but also the coal fires are emitting so much CO_2 that they were once estimated as being responsible for 2 to 3 per cent of the annual worldwide increase in carbon dioxide in the atmosphere. These figures have since been substantially adjusted downwards.

Zhang Xiangmin, a slight, young Chinese geologist, is conducting research on ways to detect coal fires using satellite images. It is a large project being run jointly by the International Institute for

Cave dwellings in loess, Xinjiang, China.

Geo-Information Science and Earth Observation (ITC) in Enschede, the Netherlands, and various Chinese institutes. I am supervising him in a team of Chinese and ITC geologists.

The problem goes back a long time. A special coal fire-fighting service was set up as early as 1958, its deputy director Zhang Lun tells me in his office in Ürümqi, the capital of Xinjiang. Ürümqi is a city with millions of inhabitants in which the decorative buildings of the original Uyghur Muslim population are gradually being replaced by the traditional architecture of the Han Chinese. The city is still bilingual, however, and the signs on the headquarters of the coal fire-fighting service are in both Chinese and Uyghur, a Turkish language written in Arabic script.

Xinjiang has nine large coalfields, Zhang Lun tells me, and there are at least 44 active coal fires. The ITC team discovered more than 150 of these using heat-sensitive satellite images. A total of 70 million tons of coal go up in smoke every year, three times the annual production of Xinjiang. There is a large model in the headquarters with small red lights showing the locations of the coal fires.

The problem is not easy to solve. According to Zhang Xiangmin's radiometer, low-grade coal layers on south-facing slopes can easily reach 75°C after being exposed to the burning desert sun all day. This is the temperature at which coal spontaneously combusts. If the fire is not extinguished, it will work its way underground through the coal seam, undermining layers above it until they collapse. This has led to whole villages subsiding. To make things even worse, the subsidence can allow air to penetrate deep into the collapsed mass of rock,

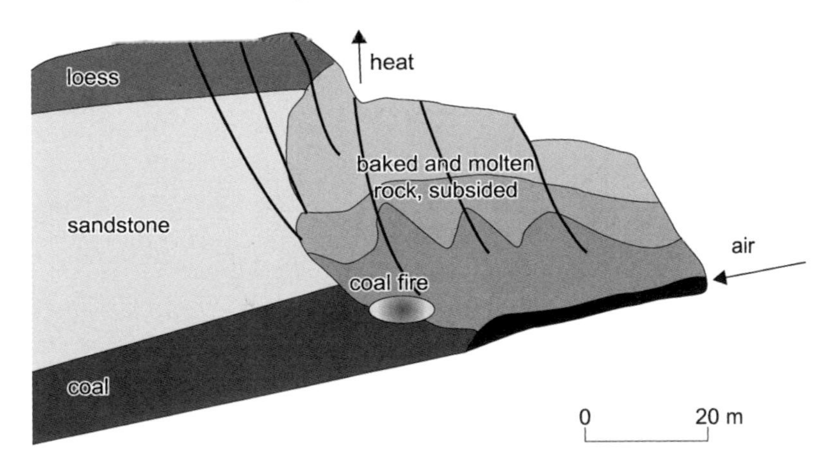

Subsiding layers and air induction above burning coal layers.

Sulfur, tar and salmiak (ammonium chloride) around vapour holes from a coal fire, Kelazha, Xinjiang, China.

creating a kind of chimney effect, so that the fire can spread even further underground. In some places there are fires up to 200 metres below the surface, where temperatures can rise as high as 2,000°C. Small lava flows of molten sandstone then drip from the rock walls. Where these natural chimneys emerge from the ground, pitch, sulfur and ammonium chloride are deposited by the hot vapours. They look almost like volcanoes. Zhang Xiangmin's research has shown that these coal fires may have been burning for a million years.

Mining only exacerbates the problem, as the fine particles that are created by mining activities can easily spontaneously combust. During the Japanese occupation they tried to solve the problem by bricking up the passages in the hope of closing off the supply of air. If there happened to be any miners still working deep in the mine, that was their hard luck. After all, reports the English-language *China Daily*, they were only Chinese. These days, the Chinese try to seal off the air holes with loess and smother surface fires with thick layers of loess or extinguish them with river water. But the fires are sometimes so deep that there is little they can do.

For the research we stay in the aptly named village of Liuhuanggou, which means 'Sulfur Valley'. We have simple lodgings above ground. The

Small dripstone cave with mini salmiak stalactites and stalagmites in the vent of an underground coal fire, Kelazha, Xinjiang, China.

old landlord was exiled there in the 1970s because he was a supporter of the Kuomintang. There is no running water here; it is brought from a reservoir every few days in a barrel on the back of a cart. At night we walk to a coal fire to look at the mountain glowing from the inside. Sometimes the flames flare up.

As the steel plant in the village has to operate around the clock, the mining continues even where the coal is burning: old bikes, broken woks, anything that can be melted down goes into the furnace, but there must be coal to keep it burning.

A miner descends a 20-metre deep vertical mineshaft in an empty barrel. At the bottom he has to fill the barrel with coal, then stand on top of it and give the signal that he can be pulled back up to the surface. We watch as the miner comes back up covered in sweat, and not only from the hard work. Zhang Xiangmin goes down in the barrel himself and measures the temperature at 20 metres with his radiometer: 70°C. Under normal circumstances such temperatures are reached only at more than 2 kilometres deep in the Earth's crust. A pump blows air into the shaft from outside, but it doesn't help: the heat down there is intense, because just below it the coal is burning. It could not be more like hell.

Coal mining above a coal fire in Xinjiang.

Later the Uyghur miner will go back to his cave-dwelling without running water, assuming it hasn't collapsed due to the coal fire burning beneath it. He cleans himself in the steam springs, a sort of sauna built on top of the coal fires. It is a simple concrete building with long, narrow strips of cloth hanging in the door opening, like wish notes for the Prophet. Inside there are large square holes in the floor through which the sulfur vapours rise freely. After an hour in there, you feel reborn, like in the Solfatara. But they keep their clothes on. There are no chairs or benches; you just have to lie on the cement floor. Now I understand why, according to Chinese Tao mythology, there are at least eighteen hells.

Since prehistoric times no one has lived underground out of choice: the Cappadocian monks fled underground to evade the Arabian and Persian invaders, the Neapolitans to escape the bombing in the Second World War, the Chinese on the loess plateau because they had run out of building materials, and the Uyghurs of Xinjiang to mine the coal. For all of them hell was outside, on the surface, and it was safer underground.

What terrestrial hell caused modern architects to design underground houses? In the village of Vals, in Graubünden, Switzerland,

Gustave Doré, the City of Dis, 1857.

Bjarne Mastenbroek and Christian Müller from the architecture firms SEARCH and CMA have designed a luxury bunker dug into a steep hillside, because the local people complained that the beautiful, virgin countryside was slowly being spoiled by architects with their newfangled designs. The Valsertal is renowned for its thermal baths and, to protect the surrounding landscape, a life-size wooden model

Steam springs and sauna above the coal fires in Xinjiang, China.

of each new building has to be made first to see if it fits in with the surroundings. In this case hell is comprised of the local villagers, planning committees and draconic building regulations. But, since you can't build a model of an underground house, the architects were granted permission to build it. The house is full of Dutch design, and offers a fantastic view of the mountains.

Villa Vals, Switzerland.

There is, however, no view of the underground, which is a pity, because it is also beautiful. Beneath the moraine layer from the last ice age, in which the house was built, are the Bündnerschiefer, beautiful glittering schists, while above, higher up on the slope, there are greenstone and limestone: the remains of the Tethys Ocean, which used to be 400 kilometres wide here, compressed and overthrusted from the depths. But there is none of this to be seen in the house. The architects wanted to look outward, not inward. And the irony is that no one lives there. You can only rent it as a holiday home. Even with so much luxury, no one wants to live under the ground permanently.

Obsidian has useful sharp edges. Landmannalaugar, Iceland.

NINE

Avarice

More shades were here than anywhere above,
and from both sides, to the sounds of their screams,
straining their chests, they rolled enormous weights.

And when they met and clashed against each other
they turned to push the other way, one side
screaming, 'Why hoard?', the other side, 'Why waste?'
. . .
It was squandering and hoarding that have robbed them
of the lovely world, and got them in this brawl:
I will not waste choice words describing it!

Dante Alighieri, *Inferno*, VII:25–30; 58–60

It seemed like a fantastic idea to Stone Age man. He was no longer an ape who happened to pick up a random stone to crack a nut. No, he needed a special stone. One the right shape to use as a hand axe, for example, to crush the skull of his prey. He found such stones in river beds: a little flattened, a little elongated, and exactly the right size to fit in your hand. There were thousands of them in the gravel banks. No one would miss them, these pebbles; that would only happen thousands of years later when his descendants would need them to make concrete or to decorate their gardens.

He discovered that if you chip pieces off some stones, you can use them to cut meat or scrape hides. Flint and obsidian were particularly good for this, but they were not easy to find in the form of pebbles: he had to dig them out of the mountain. Flint is found in clumps in limestone, and obsidian is volcanic glass. As early as the Palaeolithic era, the early Stone Age, he would walk distances of more than 100 kilometres to find the right stones, or to buy them from others.

In the Neolithic era (later Stone Age) surface supplies of stones and rocks were insufficient, and he had to start digging. This was the start of mining. He excavated flint in shafts and underground passages. He knew exactly which stones were the best and in which layers of limestone he could find them. Today the prehistoric mines would score well in terms of sustainability: disused passages were filled with mining waste and blunt tools were also left in the mine, so that there must have been little visible waste outside, on the surface.

But there is a downside. There are now large, ugly holes in the ground, which are still visible after 5,000 years. The Stone Age miners had no Turkish slippers, so we will never know what the excavated layers looked like in their virgin state. Important evidence may have been lost to science forever.

The amazing thing is that, 5,000 years ago, flint was the useful resource and limestone was seen as waste, while today limestone is a very sought-after mineral for the cement industry and flint is waste, or at the most a by-product used to make durable white stripes on roads.

I once conducted research with Leiden archaeologist Aad Boomert in Suriname into the rocks that Indians used in the pre-Columbian era

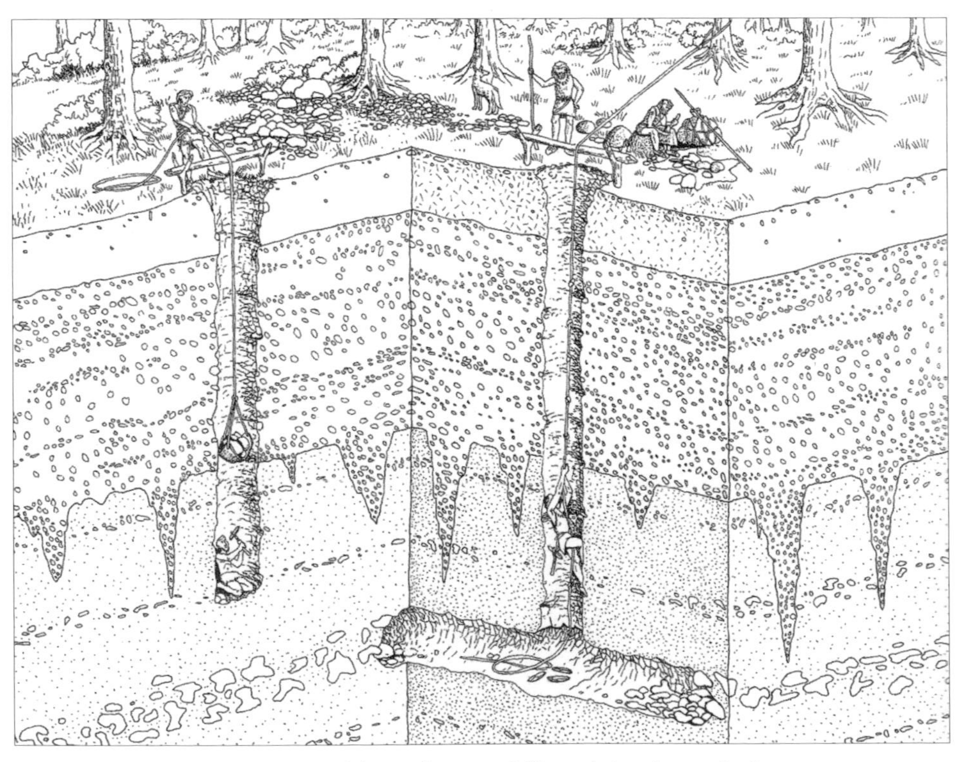

Reconstruction of the underground flint mining that took place in Rijckholt 5,000 years ago.

to make stone axes. We examined more than 800 stone tools from archaeological excavations of various Stone Age cultures: two-thirds consisted of stone axes and the rest were grindstones and mortars (*metates* and *manos*), polishing stones, arrow shaft polishers and a few small amulets. The results were striking: they used a different stone for each kind of tool. For axes they preferred metabasalt, a fine green stone that is easy to sharpen on a rock in the river, using water and sand, but which does not split or break easily. There are not many places where metabasalt occurs naturally and, at one of the most important sites at Brownsberg, a workplace has been found where rough prefabricates were made. Stone axes of metabasalt have been found in archaeological excavations as far as 800 kilometres away. For grindstones they preferred to use granite, which has a hard, rough surface due to protruding pieces of quartz. The Stone Age people were good geologists.

They also used mineral pigments for their rock drawings, including yellow, brown and red ochre and black manganese oxide. The creators of the famous Lascaux rock drawings, which date back 17,000 years, probably walked at least 40 kilometres to find hematite for the red colours. The oldest mines in the world were also pigment mines: red hematite was excavated 40,000 years ago in the Lion Cave in Swaziland – interestingly enough, from rock that was also among the oldest in the world: iron deposits between 3 and 4 billion years old.

You are dependent on nature, you are hunting, and you look everywhere around you for something that may be useful – wild animals, wood, fruit – and then suddenly you see a beautifully coloured stone: blue, green, turquoise. Or you are fishing and you see something glinting in the water. When you pull it out, it doesn't go dull, like all the other stones, but continues to glisten. You take it home with you and get a beaming smile by way of thank you for the lovely gift. Then the woman next door wants one, too. Word gets around and before you know it, you are no longer a hunter, but a miner. And you become rich: wealth through possession, something that distinguishes people from other animals – unfortunately, you might say.

The oldest known jewellery is made of shells of ivory, but copper beads 10,000 years old have been found in excavations in Aşıklı Höyük, Turkey, which show traces of being produced using heat, 5,000 years before the Bronze Age. The Chinese found stones of nephrite jade 8,000 years ago in rivers in the Kunlun Mountains near Khotan in Xinjiang, and the Egyptians excavated turquoise from mines in the

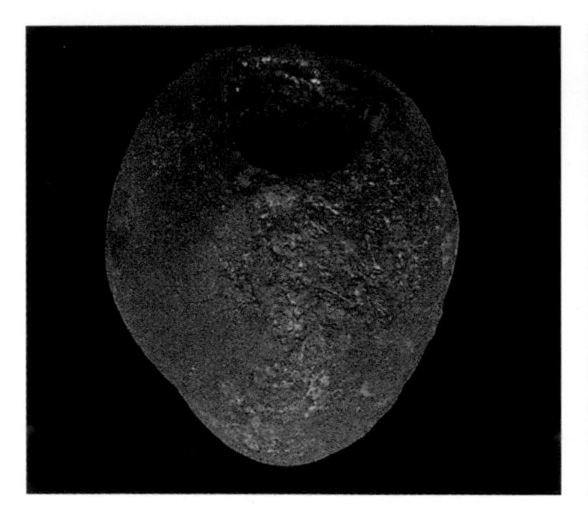

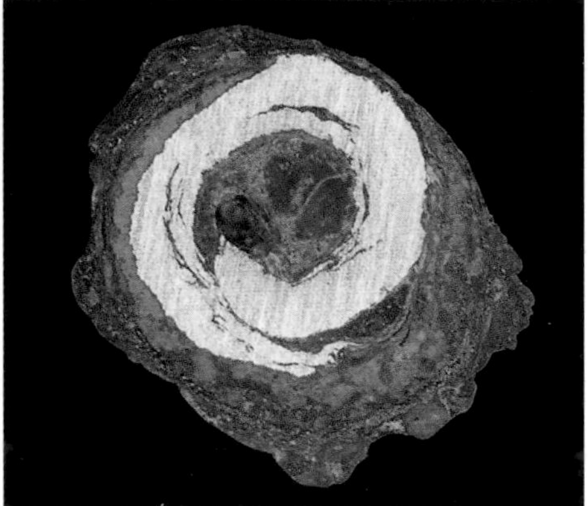

A 10,000-year-old copper bead (left), Aşıklı Höyük, Turkey. The cross-section of a bead (right), metallographic image: an early gobstopper.

Sinai 6,000 years ago to use in their scarabs and other jewellery. British archaeologist Peter Roger Stuart Moorey specifies 63 different types of stones used by the Mesopotamians, not only for building, but also many precious and semi-precious stones for seals, pendants, amulets and beads. According to historical scientist Robert J. Forbes, the Sumerians listed more than 180 minerals known to them according to external features, 120 of which were used in medical and magical texts. 'Generally speaking, their ranking is astoundingly good', he concludes.

Lapis lazuli in particular was highly valued. In the Sumerian epic of Gilgamesh the lovely goddess of fertility Inanna had to remove all her lapis lazuli beads during her descent into the Underworld. Archaeologists found more than 6,000 objects of lapis lazuli in the burial chambers at Ur. But the closest lapis mines are 1,500 kilometres away, in the province of Badakhshan, in Afghanistan. Marco Polo said that this 'azzurro' was of 'the finest quality in the world', when he visited the mines around 1272. Globalization *avant la lettre*.

A fantastic source of information on stones in antiquity is the book *On Stones* by the Greek Theophrastus, a pupil of Aristotle who lived from approximately 371 to 287 BCE. It is one of the first books on geology to have been preserved; only the Chinese *Shan Hai Jing* (Classic of the Mountains and Seas) from the fifth century BCE is 200 years older. Much of what later authors, such as Pliny the Elder, write about rocks comes directly from Theophrastus. He describes the properties of the main types of 'stone' in detail: precious and ornamental stones used for jewellery, amulets and seals, and the 'earths', mainly pigments and dyes, also saying where they can be found and mined.

Emerald, sapphire, onyx, jasper, chrysocolla, hematite, agate, cinnabar, gypsum: they are all names that are familiar to us today.

But if you read *On Stones* closely, almost none of the descriptions still fit the current names. For example, according to Theophrastus, sapphire has golden flecks in it, a typical property of what we now call lapis lazuli. Only the blue colour is the same. *Smaragdus* is green, but is not emerald, while gypsum is not only what we now call gypsum, but also calcium oxide, which was used to produce mortar for building. The way the names of stones have shifted throughout history is told in an inimitable fashion by German linguist Hans Lüschen in his book *Die Namen der Steine*. Of the twelve precious stones mentioned in the Revelation of St John, only three still have the same name: hyacinth is our sapphire, their sapphire was our lapis lazuli, and our hyacinth is a kind of zircon. It is the same story as with the names of the rivers.

It is hardly conceivable that Stone Age man had no knowledge of metals: he must have seen chunks of gold glistening in the river, but as yet did not possess the techniques required to do anything with it. In effect the whole division of prehistory into the Stone, Bronze and Iron Ages is based on the use of materials acquired by mining. Until the Middle Ages gold, silver, copper, iron, lead, mercury, tin and bismuth were the only known metals.

And suddenly you see a beautifully coloured stone.
Copper ore, Gruvberget, Sweden.

The copper mines of Timna in the Negev desert ('a land whose stones are iron, and out of whose hills thou mayest dig brass', Deuteronomy 8:9) were already in use 6,000 years ago. At first copper clumps were collected on the surface and taken to a work camp, where they were crushed in large stone mortars to remove the richest ore and melt it in primitive ovens. Later miners used heavy stone hammers on the Pillars of Solomon, jagged vertical white sandstone cliffs, to get at the ore. Analyses of the ore they used shows that they already had an excellent eye for the amount of copper in the rock. But the tools of the copper miners were still made of stone. In the late Bronze Age/ early Iron Age, copper mining was already a large-scale enterprise. The end product was copper, not yet bronze. For bronze you also need tin, and copper and tin are rarely found together in ore deposits. That is why archaeologists are very keen also to discover the origins of the tin.

Near Kestel in Cappadocia, on the Anatolian plateau, Turkish archaeologists have found very old mineshafts which they suspect were used to excavate tin. The shafts were made by starting a fire close to the rock face. The unequal heating of the minerals in the rock created such tension that the rocks would break off violently. This technique, known as fireblasting, is still applied, using flamethrowers. Ore is often found in hard rock, so simply digging, as in tuff or loess, is out of the question. Carbon dating of charcoal found in the passages shows that the mine was used in 2000 to 2800 BCE at the earliest. But there is no trace of tin ore. The old miners depleted the ore deposits so thoroughly that until 1995 it was not determined beyond doubt that Kestel had indeed been a tin mine. We will never know how the ore was located in the rock before they removed it. It is a shame they didn't leave just a minute amount, if only to help those who tried later to understand their secrets.

Gold jewellery was already known from the fifth millennium BCE, in a broad band stretching from Bulgaria to Egypt. There were gold mines everywhere in Upper Egypt, sometimes with shafts as deep as 50 metres. Gold-panning was also a thriving industry, according to the stele of Sa Hathor, who lived in the time of Pharaoh Amenemhat II, around 2400 BCE.

Ancient Athens owed much of its wealth and power to the silver mines in Laurion, in the most southeasterly point of the Attican peninsula. The silver and lead ore is concentrated in two veins in a series of folded marbles and schists dating from the Triassic and Jurassic

Tin mine from the early Bronze Age near Kestel, Turkey.

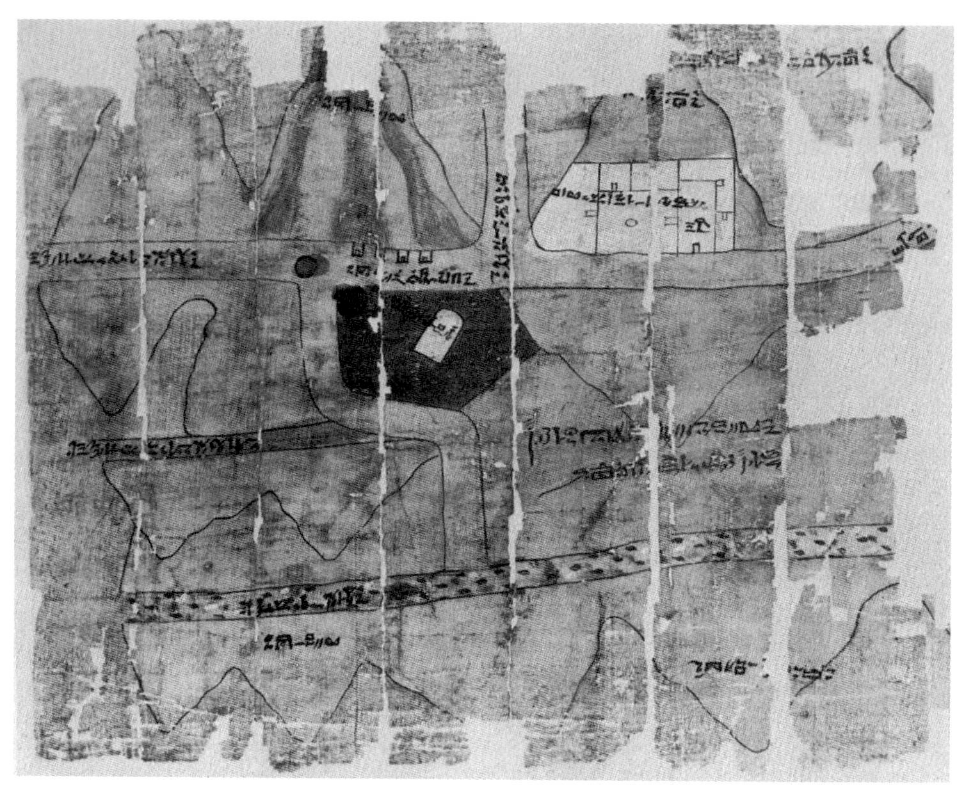

The Turin Papyrus, an ancient Egyptian map of a gold mine, *c.* 1160 BCE.

era. There are indications that they were used as early as the Minoan period, around 1500 BCE. From 500 BCE, according to Aristotle, the income from the mines was included in the Athenian budget. There were more than 500 pits in the area, and many famous Greeks, including Thucydides and Demosthenes, owned shares in them. During their heyday between 10,000 and 20,000 slaves worked in the mines, hacking and digging for up to ten hours at a time, squatting, kneeling or even lying down. The mine builders dug shafts up to 100 metres deep, a formidable achievement given that modern speleologists were unable to stay longer than half an hour at a depth of 30 metres without using oxygen tanks. Pliny reports that the heavy fumes in the mines were especially bad for dogs: another Grotta del Cane.

Digging underground was a hellish activity, and not only in Laurion; even Sisyphus was better off in the Underworld with his stone. Theophrastus writes that 'it is not possible to stand upright while digging in the pits of Samos' (where they were mining kaolin for making pottery) 'but a man has to lie on his back or on his side.

The Turin Papyrus map according to William Gowland, 1912. (A) Gold-panning area; (B) Gold-bearing mountains; (H) Building for storing gold; (S) Stele for Pharaoh Seti I; (T) Pond or tank. The large building is a temple to Ammon.

Greek miners.

The vein is about two feet in height, though much greater in depth.' Dante could not have dreamed it up. Elsewhere he says that in the ochre mines in Cappadocia 'the risk of suffocation is a serious matter for the miners, since this can happen to them quickly'. He is probably referring here to the mercury mine in Iconium (Sızma in the Konya Basin), which was already used by the Hittites and the Phrygians around 1500 BCE. The red mercury ore cinnabar was also used as pigment, and mercury vapour can quickly be lethal in large concentrations. During archaeological excavations in this mine, 50 skeletons were found, perhaps the bodies of people trapped in the mine after the sudden collapse of a passage, or suffocated by mercury fumes. Robert Forbes, who we have come across before, writes that in antiquity mines rarely went deeper than 100 metres, and exceptionally to 300 metres, because of problems with groundwater and drainage. Lighting and ventilation were, of course, also problematic. In the fifteenth-century silver mines of Ste Marie-aux-Mines in the Vosges region of France, children had to run back and forth through the narrow tunnels the whole day to ensure sufficient ventilation.

There is good reason why Lucretius in *De rerum natura*, in the first century BCE, referred to mines as 'birdless places' (Scaptensula was a mine in Thrace):

> See you not, when they are following up the veins of silver
> and gold and searching with the pick quite into the bowels
> of the earth,
> what stenches Scaptensula exhales from below?
> Then what mischief do goldmines exhale!
> To what state do they reduce men's faces and what a
> complexion they produce!
> Know you not by sight or hearsay how they commonly
> perish in a short time and how all vital power fails those
> whom the hard compulsion of necessity confines in such
> an employment?
> All such exhalations then the earth steams forth
> and breathes out into the open air and light of heaven.
> (VI:808–17)

Were the mineworkers not afraid of the Underworld? Darkness, harmful vapours, strange sounds, rolling stones like Sisyphus? Wouldn't they eventually dig through the roof of the Elysian Fields or, even worse, Tartarus? This thought did not seem to have occurred to any of the classical writers. But there was a lot of magic surrounding the mines.

Some of this magic even remains in the monumental treatise on mining *De re metallica* by Georgius Agricola (Georg Bauer), published in 1556. This book is the standard work, the bible for all mining engineers. There is an excellent 1912 English translation by former president of United States Herbert Hoover, who was originally a mining engineer, and his wife Lou, a classicist. The Hoovers not only translated the text but also added a biography of Agricola, reading so much literature on the subject and commenting on it in the footnotes that the book remains an unrivalled resource even a century later. The original sixteenth-century engravings it contains have become classics.

Agricola begins his book with a passionate plea for mining that is at times very modern. A miner must know where you can find ore, what kind of ore it is, and if it is in veins or lodes. He must know how to construct a mine and how to analyse ore. And he must understand the 'philosophy' – we would say the science – of the origins, causes and nature of the underground, and of health and medicine, to protect his miners and other workers. He must have some understanding of astronomy to measure the orientation of the ore-bearing rock, and he must understand levelling to know how deep to dig a tunnel. He must be able to calculate so as to determine the costs of his equipment and of the mining activities. He needs to be an architect so he can design and assemble his own machinery and underground timber constructions, and draw building plans. And of course he must know the law, so that he can make legitimate claims and not start mining on other people's land.

Many people are against mining, Agricola says. Some believe that farming offers a more stable income but, he says, the silver mines in Freiberg have been working for 400 years, the lead mines in Goslar for 600 years and the gold and silver mines in Chemnitz, where he himself lived, for 800 years. Some complain that farmland is destroyed, forests cut down, and rivers and streams polluted by toxic waste, causing more damage than the benefits of mining. And they say that we don't need metals at all, as riches from gold and silver only lead to jealousy, conflict and war. Blessed are those, like Socrates, who care not for property.

Agricola responds by saying that if a tyrant falls in love with a beautiful woman and besieges her city, it is not the fault of the woman's beauty, but of the tyrant's unbridled lust. His opponents claim that

How to descend into a mine, according to Agricola in his *De re metallica* (1556).

iron is used only to make weapons, but it is not the fault of the iron that people want to kill each other: that is the evil nature of man. Iron is also used to make ploughs and sickles that farmers use to grow grain, knives to cut meat and hooks to catch fish; man can no longer live without metal. And most mining takes place in mountainous regions where there is little fertile ground, while the income that the mines generate can be used to buy fish and meat elsewhere.

Others say that the Creator hid metal ore inside the Earth for a reason: that he did not intend it to be removed, as metal brings only misery. But then they also criticize the Creator, says Agricola, as that would mean he intentionally created things that have no use or which are evil. The Earth does not contain ore because it does not want man to remove it, but because nature, which is providential and wise, has given everything its place. It placed ore in rock, because the ingredients it requires to exist cannot be found elsewhere. If ore were formed in the air, it would fall to the ground.

In the remainder of the book Agricola describes in detail the characteristics of miners, the search for ore, the classification of different ores and rock, and all the techniques that are required for mining. It is a very readable text, written by someone who was both a scientist and an engineer, and who was familiar with and, where necessary, commented on, the historical works of Theophrastus, Pliny and many others.

It therefore comes as a great surprise when you suddenly read in Book VI: 'In some of our mines, however, though in very few, there are other pernicious pests. These are demons of ferocious aspect, about which I have spoken in my book *De Animantibus Subterraneis*. Demons of this kind are expelled and put to flight by prayer and fasting.' In a footnote, the Hoovers quote what Agricola says in his other book: 'Then there are the gentle kind which the Germans as well as the Greeks call *cobalos*, because they mimic men. They appear to laugh with glee and pretend to do much, but really do nothing.' Then there are *trulli* and *suions*, which take on the form of people. But, Agricola writes a little later: 'The fifth cause [of evil] are the fierce and murderous demons, for if they cannot be expelled, no one escapes from them.'

What else did the demons do? They played tricks on the miners. They made stones glitter, even though they contained no gold, silver or lead. Miners were blinded by the stones and called them 'blende'. 'Blende is a glittering stone, black and sometimes yellow, that contains no metal and often blinds and misleads the miner', wrote Johann

Heinrich Zedler in 1733 in his *Great Encyclopedia*. Blende usually refers to zinc blende, zinc ore, but zinc had not yet been discovered as a metal. Miners did not learn how to extract zinc until later in the eighteenth century. Agricola's *cobalos*, *Kobolde* in German, teased the mining engineers in the same way, and gave their name to a glittering stone that was of no value. 'Cobalt is a toxic, predatory, malignant kind of ore, that dilutes the good ore, or makes it wild and cold, and many believe that it steals the silver', wrote Johann Hübner in 1727 in his mining lexicon. A *Nickel* was another subterranean demon that tried to hide its true nature (*necken* means to torment). Cobalt and nickel are now very sought-after metals.

There are countless legends about mountain spirits. Many medieval miners were poor wretches, pioneers who often tried their luck in the mines out of pure misery, as a last resort to support their usually large families. Finding a rich vein could be a matter of life and death. They were often unskilled and had no idea of where ore could be found or how it was formed, and therefore how to look for it. They knew nothing of Agricola. Success was largely a matter of trial and error. Ore-bearing rock can be so capricious that even now it is difficult to give general instructions on how to find it. No wonder that the complete unpredictability of good or bad luck, the darkness under the ground, the bad air, the scary noises, the sudden flows of water and the rivalry between the miners provided fertile ground for rumours, stories, fantasies and legends. Good examples are the *Berggeist* (mountain spirit) of the Ore Mountains, or the Old Man and Kaspar in tales about the coal mines in the Dutch province of Limburg. But not all mountain spirits sought to do harm. The little men would often show the poor fortune-hunters a rich vein, but always with conditions attached: they were not allowed to tell anyone else, or they had to promise to give the spirit something valuable. If they boasted about the find to their friends in the pub, or tried to keep them away from it, the next day the vein would once again be *taube Stein*, worthless rock.

It is the same story in the Potosí silver mines in Bolivia. The Cerro Rico still looks like a volcano, even though it was extinguished 14 million years ago. It no longer has a crater; the holes in this mountain are on the sides. If you look closely in the hard Andean light at an altitude of 4,000 metres, you can see that it is peppered with small, square holes: mineshafts. It has been hollowed out more than practically any other mountain on Earth. Silver has been mined here for 500 years; it was the richest silver mine ever discovered by the Spanish conquistadors

Potosí, Cerro Rico, Bolivia.

and was for a long time the largest in the world. Besides silver, it also contains a lot of tin, lead, zinc, wolfram and bismuth. The Incas never mined here because, according to a legend from that period, in the words of the Spanish Jesuit José de Acosta in his *Historia natural y moral de las Indias* from 1590, 'the Indians heard voices telling them not to touch anything there, that the mountain was reserved for others'. The first vein was registered on 21 April 1545 by the Spaniard Juan de Villaroel and the Indian Guanca, who had discovered it. The rich silver vein 'lay above the ground the height of a Lance, like unto Rocks, raysing the superficies of the Earth, like unto a crest of three hundred foot long, and thirteene foot broad, and that this remayned bare and uncovered by the deluge'. Potosí was for the Spanish what Laurion had been for the Athenians. Silver coins were suddenly three times thicker than before the *conquista*.

Yet the Cerro Rico, too, has its demon: el Tío, the Uncle. There are many stories about el Tío. In *Cuentos mineros del Siglo XX*, Bolivian writer Luis Heredia Heredia tells the tale of the poor wretch Ciriaco Limachi. He was always afraid in the mine, dared not work the night shift, and was so poverty-stricken that he had no money even to buy coca leaves, and fell deeper and deeper into debt. One stormy night he decided to risk working a double shift. He made a pact with el Tío and found so much silver that he became immensely rich and bought the whole mine. He used to light his cigarettes with banknotes and started drinking, always raising his first glass to his comrade, el Tío. On another stormy night, he went into the mine drunk. The last man who saw him found him praying in front of the Cristo del Socavón, the Christ of the Mineshaft, but no one ever saw him again. After that, his rich vein suddenly became sterile.

Today's miners are still poor wretches, and el Tío is still there. If I want to enter the Cerro Rico, I have to make offerings. There are dozens of stalls in front of the entrance to the mine selling dynamite and coca for the miners and strong drink for el Tío. I follow the guide deep into the bowels of the extinct volcano. Since the tunnels have been hacked out at Inca height, I have to bend forward and find it difficult to keep up with the guide's fast pace. Every now and then, I bang my helmet against a protruding point of rock in the ceiling of the passage. It has an old-fashioned carbide lamp and, as I hobble on behind the guide, I hold the pack of dynamite out in front of me, as far as possible from the flame. Perspiring men pull mine trucks full of ore, dented and with peeling paint, along the narrow rails. A boy of nine or ten carries half a car tyre full of stones up a rickety ladder to the gallery above us. Deep

Dynamite (left) and coca leaves (right) for the miners.

A miner visiting el Tío, Potosí silver mine, Bolivia.

in the mine, in a niche in the wall, sits el Tío. He has a frightening black ape-face, the corners of his mouth turned down, glowing eyes and devil's horns on his head. The confetti around him ought to soften the shock . . . I quickly make my offerings.

The mine should have closed long ago. It is only kept open under pressure from the trade unions. Miners are proud people. There is not much silver left in the mine, and the miners now mostly excavate tin and lead. Their average life expectancy is 45. A recent novel about the lives of the miners has the title *Socavones de angustia*, the Shafts of Fear. For those who work in the hell of Cerro Rico every day, the hereafter must come as a respite.

The small mountain spirits in the mines ruled only over their own ore mountain, and were not part of a universal hell. It did not occur to the tortured miners to associate them with God's wrath. Nevertheless they still needed the protection of the saints. In the mines of Saxony Jesus himself protected the gold against evil spirits, Mary the silver and St Anna the tin: *vive la différence*. St Wolfgang was a favourite because he could determine the size of a vein of ore with a single throw of an axe. St Daniel sought ore in the trees, in the same way that Aeneas had to pick a golden branch for Persephone before he was allowed to enter the Underworld. And anyone who had prayed to St Christopher before going into the mine did not need to be given the last rites. But most venerated of all is Barbara, the patron saint of miners: she was even asked to give her blessing to the digging of a new metro tunnel in Amsterdam in 2010.

But you must not give the saints a free hand, either. That is clear from the story of the underground salt cathedral at Zipaquirá in the Cordillera Oriental in Colombia. The Muisca people mined salt here before the *conquista*, transporting it in large blocks over enormous distances and trading it for gold and emeralds. When Don Jiménez de Quesada, on the orders of the governor of Santa Marta, took 800 Spanish soldiers to look for the sources of the large river Magdalena in 1537, he met Muisca salt traders. They made such an impression on him that, instead of following the river, he deviated from his route and found himself on an broad plateau at an altitude of 2,600 metres in the Cordillera Oriental. It was a severe journey, during which he lost most of his soldiers, and when he was attacked by the Muiscas near the salt mines, he withdrew to the south and, in 1538, founded what would later be the city of Bogotá, the capital of Colombia. The Spanish later gained control of the mines. The capital therefore owes its location to the salt of Zipaquirá.

In 1801 the German explorer Alexander von Humboldt visited the mines and proposed excavating the salt by digging shafts instead of the customary opencast mining. The salt, the remains of an evaporated sea from the Cretaceous period, does not form a salt dome, as in many other areas, but a series of layers, tilted towards the west and pressed between layers of black shale. Digging shafts, the miners penetrated deeper and deeper into the mountain and the network of tunnels now covers more than 800,000 square metres. The patron saint was Our Lady of the Rosary of Guasá and, from the moment that the mine was opened to the public in 1923, she became such a popular object of worship that, in 1953, the mine was officially consecrated as a cathedral.

This was the cathedral I visited in 1980 when I was living in Colombia, and I was struck by the beautiful domes that had been cut from the salt in Spanish Baroque style. The regular layers of thick, glistening salt and thin, dark bands of shale were a magnificent sight. In 2008 I visited the cathedral again and was surprised to see that the entrance looked nothing like it had in 1980. The Baroque domes had

La Virgen del Rosario de Guasá, Zipaquirá, Colombia. According to legend, the miners scratched this niche out of the salt with their nails. The thick white layers are salt, the thin black ones are shale.

disappeared and everything was modern, with rigid lines. It was much more spacious and in the back, under the largest vault, an enormous cross had been hacked out of the salt, illuminated in the semi-darkness by hidden green lights. What had happened here? Had they, like the cave dwellers of Cappadocia, simply expanded and remodelled the old chambers?

I found the answer in the book *El milagro de la sal* by Gustavo Castro Caycedo, brother of the renowned Colombian journalist and

The second salt cathedral at Zipaquirá.

writer Germán Castro Caycedo, born in Zipaquirá. In the time that I lived there Germán's book *Colombia amarga* was the most impressive testimony to the country's violent history. *El milagro de la sal*, the miracle of the salt, is less critical, less subversive and somewhat chaotic. Gustavo did not have his brother's talent. But I found what I was looking for: the cathedral I had visited in 1980 was no longer there. The old 100-metre-deep mine had become unstable and the roof was in danger of collapsing, so the cathedral had been closed in 1991. A new cathedral was hacked out of a salt layer 60 metres deeper than the previous one, where miners had already excavated salt eleven years earlier. Now it can accommodate 10,000 people, concerts are held there, and it has become one of the country's greatest tourist attractions. Castro Caycedo could not resist including jubilant praise for the cathedral from the foreign press and tourist brochures in his book: not only in Spanish, but also in English, German, French, Italian, Portuguese, Chinese, Korean and even Basque. That's one way to fill up your book.

Perfect irony! Science, in the form of Alexander von Humboldt, advises digging in the ground, religion takes possession of the resulting hole, but nature strikes back! It is not hell punishing those who violate its domain from below, but nature making the roof collapse from above. Religion responds by digging deeper, even closer to hell, but sooner or later, what Degryse and Lévy-Leblond predicted for Dante's Inferno will come to pass: the roof will no longer be strong enough to support the large hollow space. And who did Dante have in mind in the fourth circle when he said 'It was squandering and hoarding that have robbed them of the lovely world'?

> The ones with bald spots on their heads
> were priests and popes and cardinals, in whom
> avarice is most likely to prevail. (VII:46–8)

No one knows better that the roof of a mine can collapse than the old miners of Falun in Sweden. Falun is the oldest and most famous copper mine in Northern Europe. It has certainly been in use since 1300, and perhaps even since the year 1000. In its heyday, from the seventeenth to the nineteenth century, 70 per cent of the copper used in Europe came from Falun. The roof of the Palace of Versailles is made of Falun copper.

Alongside my study of physical geography at Amsterdam, I did a minor in petrology and mineralogy under Professor De Roever, later my PhD supervisor, and Professor Oen Ing Soen, an Indonesian-born mineralogist who had studied the Bergslagen region around Falun for twenty years. Both subjects satisfied my hunger for beautiful minerals and rocks. I would sit hunched over the microscope in the practical room for hours identifying thin sections of rock. You get a hand specimen, too, a grubby piece of rock as large as a fist and greasy from generations of students' fingers. It is so worn, you can hardly see the miniscule grains of ore mineral.

The high point of the minor was an excursion to the Swedish and Norwegian mining areas in 1970: zinc from the Zinkgruvan at Åmmeberg, iron from Taberg and Grängesberg. Grängesberg! That was the deepest point that we reached, 400 metres below ground. I was impressed by the fast descending lifts, the spacious chambers, the drilling technology, the beautiful rocks, but perhaps most of all, by the small garden that the mining company had laid out down there in the depths, at the entrance to their office. A few square metres of greenery, 400 metres underground, which – under the strong lights – actually seemed to be growing, too. A miniature Elysian field.

And then, in Dalarna, the renowned copper mine at Falun, Stora Kopparberg. There in Falun we could see glittering, metres-thick veins of copper shooting through the blasted open rock walls. They are classic rocks, 2 billion years old, and they can be found in every textbook. And all those rare minerals that we knew only as small chunks in cardboard trays in the wooden rock drawers in the practical room – you could pick them up here in the waste heaps from the mine. And a sample had to be of a decent size. If De Roever considered one to be too small, he would say contemptuously: 'That is a *Damenhandstück*!', an expression he had learned from his Swiss mentor Ernst Niggli. We gathered tens of kilograms of stones: it was a miracle that the axles of the bus didn't succumb under the weight.

In Falun we descended by lift, helmets and raincoats on, to an old mineshaft 60 metres deep. Quite luxurious, if you think how they used to do it in medieval times. Then, too, they would go down seven at a time – in a wicker basket hanging on a hemp rope and lowered into the shaft with a winch. Seven men bunched into one basket – this was only possible because each one stood with one leg in the basket and the other hanging over the side, while hanging onto the rope. They could tell by the sound of the rope if it was going to hold their weight.

The famous botanist Carl Linnaeus also went down into the mine in 1734, shortly before he left for the Netherlands. He recounts what the miners had to go through:

> We went the whole way down by wooden ladders, mostly of twenty rungs each, hanging vertically and free of the wall. Often they were joined in pairs and only supported at their ends, so that they swayed about. Below the drifts were so low that we had to stoop or crawl, and so narrow that many a time we had to turn our bodies sideways in order to move forward; again and again I knocked my head on projecting bits of the roof – a roof that was covered in crystallized vitriol of a curious blackish colour. All the men carried torches in their mouths. At the bottom there blew a cold wind strong enough to turn a windmill. The horses which drew the winch were driven by a man who stood close to the axle, and there were stalls, hay and a smithy. The ore was carried in wheelbarrows or small four-wheeled waggons.
>
> In these gloomy places to which no daylight ever penetrates, these doomed creatures – there were about twelve hundred of them – lived and had their being; yet they seemed to be content,

> because they fought to get jobs there. They are surrounded on every hand by rock and gravel, by dripping corrosive vitriol, by smoke, steam, heat and dust. There is a constant risk of sudden death from the collapse of a roof, so that they can never feel safe for a single second. The great depth, the dark and the danger, made my hair stand on end with fright, and I wished for one thing only – to be back again on the surface. These wretched men live by the sweat of their brows, working naked to the waist and with a woollen rag tied over their mouths to prevent them so far as is possible from inhaling fumes and dust. The sweat poured from them like water from a bag. It was only too easy to fall into a hole, to miss one's footing on the rung of a ladder; or a rock might come crashing down and kill some miserable man instantly. Every aspect of hell was here for me to see.

In the same year that Linnaeus visited Falun, a fellow countryman, geologist and mine inspector, Emanuel Swedenborg, published a book, *Opera Philosophica et Mineralia*, in which he described the mining methods in Falun. Swedenborg did not see hell, but something completely different. In the foreword, he writes:

> Consider what great quantities of copper this mine . . . has brought forth for thousands of long years and that are yet being born from its inexhaustible womb . . . In short, the whole floor of this mine, all its walls, doors, chambers, halls and pillars, all its open surfaces gleam in all directions with this wonderful vein's glittering, golden rays. The visitor could believe he had been carried to Venus herself, when the goddess stood as a bride in an exquisitely decorated chamber, and there happily welcomed her guests from Falun.

Later in his life Swedenborg had a radical change of heart. On 6 April 1744, during a visit to Delft, he had a vision, turned his back on science and became deeply religious. He wrote a large number of theological works, the best known of which is *De caelo et ejus mirabilibus et de inferno: Ex auditis et visis* (1758), usually referred to simply as *Heaven and Hell*. He describes with scientific accuracy – that much he had retained – what heaven and hell look like, which had been 'given to him' to perceive with his own eyes.

There are countless hells, he writes in the book, all 'divided and regulated with the utmost exactness and congruity', some on top of the other, and 'communicating with one another, some by passages and some by exhalations'. There are hells 'under every mountain, hill, rock, plain and valley'. Hell is, as it were, dug out below ground and continues down to the depths. It seems to be inspired by Falun. The doomed souls in hell are 'hideous and ghastly, ugly with warts, carbuncles and running sores'. They shun the light, 'and seek to appear in their own light, which resembles that which comes from fire coals, or burning brimstone'. And what are they punished for? According to Swedenborg, the greatest sins for which you end up in hell are lust and love of self. Swedenborg, the eternal bachelor, had seemingly forgotten about the temple of Venus.

Whether it was Venus or hell, Falun tended to invoke religious associations. Hans Christian Andersen was a little more matter of fact: he compared the stench of sulfur at Falun with that of the Solfatara.

The greatest disaster at Falun actually took place half a century before Linnaeus' visit and Swedenborg's book. On 25 June 1687 the roof collapsed above four shafts: the Blankstöten, the Skeppsstöten, the Bondestöten and the Måns Nils. The miners were not yet so good at calculating the strength of the roof at that time, and they probably didn't know just how close the different tunnels had come to each other. The subterranean vaults had steadily increased in size, the remaining pillars had become thinner and, to make things worse, as Swedish mine historian F. R. Tegengren writes in *Sveriges ädlare malmer och bergverk* (1926), they had piled up large mountains of mine waste on the roof. The collapse left an enormous hole 1.5 kilometres wide and 95 metres deep. Luckily it was a free day and no one was killed. The mountain spirit of Falun, Gamla Mormor, the Old Grandma, was in generous mood that day. She was the daughter of a giantess who guarded her treasures jealously and threw stones at you if you tried to steal her ore. Selma Lagerlöf tells the story beautifully in *The Wonderful Adventures of Nils*, a geography book in disguise for Swedish schoolchildren that I devoured as a child.

One famous story is that of miner Fat-Mats Israelsson, who had been confined in the mine in 1677 when a tunnel collapsed, and was found again in 1719 after the great collapse. His fiancée Margareta Olsdotter, now a grey old woman who had always waited for him, recognized the body. He still looked as young as the day he disappeared: his body had been preserved by the sulfuric acid released by

The collapsed mine at Falun, 1968.

the oxidation of the copper ore. Linnaeus, whose wife came from Falun, saw the exhibited body and predicted that it would quickly decompose as it was only covered with a crust of 'vitriol' (sulfates). He was proved right and, in 1749, Fat-Mats was buried.

The story inspired E.T.A. Hoffmann to write his novella *Die Bergwerke zu Falun*, in which a young sailor is persuaded by a friendly old man to say farewell to the sea and work in the mines at Falun. The young man is successful and becomes engaged to the boss's daughter, but the friendly old man reveals himself underground to be an evil mountain spirit, who predicts that he will never marry his sweetheart.

On the morning of the wedding the young man goes quickly down into the mine, telling his bride: 'Down in the depths below, hidden in chlorite and mica, lies the cherry-coloured sparkling almandine, on which the tablet of our lives is graven. I have to give to you as a wedding present.' While he is down there, the mine collapses. The young man does not return, and is not found until 50 years later. He still looks the same and the only one who recognizes him is his fiancée, now old and grey, who has waited for him all those years.

In 2008 I once again went down into a Swedish mine, in Kiirunavaara, the largest iron mine in Europe. This is even more luxurious than in 1970, since we no longer need a lift. With our helmets, headlamps, safety jackets and boots on, we ride in the company bus – with a sign saying 'Welcome to the underground' – into a tunnel that dives down steeply, with a slope of 10 per cent, deeper and deeper, with no light at the end, because the end is 1,365 metres below ground. Inside this iron mountain there are 400 kilometres of asphalt roads, complete with road signs and traffic lights at the junctions. Every minute 100,000 cubic metres of fresh air is pumped into the mine. We drive on until we reach a depth of 530 metres. There is the information centre and an office where young people sit watching monitors and typing on keyboards. I search in vain for the subterranean garden of the Grängesberg mine of 40 years previously. A quarter of a million tourists come here every year, and you can even marry underground, just as in Naples in the war. It is a strange sensation; you don't feel as though there is half a kilometre of rock above your head.

We are shown into an underground cinema. Our host, Kari Niiranen, tells us about the history of the mine. The ore was discovered as long ago as 1696, but mining only started seriously around 1900. The Saami (Lapps) in the area did not want the mountain to be mined, as it was sacred to them. We know now that the vein of ore is 100 metres thick, 5 kilometres long and goes down into the ground at a steep angle for at least 2 kilometres. Two kilometres! That's why the tunnel is so deep. That is on a complete different scale to previous centuries. And the Swedes are planning to exhaust the deposits, too. In the reception hall there is a large three-dimensional Perspex model of the mine, with each working level depicted by a horizontal sheet of Perspex. I count 30 levels. There is also a vertical cross-section of the mine engraved in stone, which shows that since 1960 the miners have increased the rate at which they have penetrated deeper into the ground, at first 10 and now 20 metres a year. Measured from the surface, the ore is already

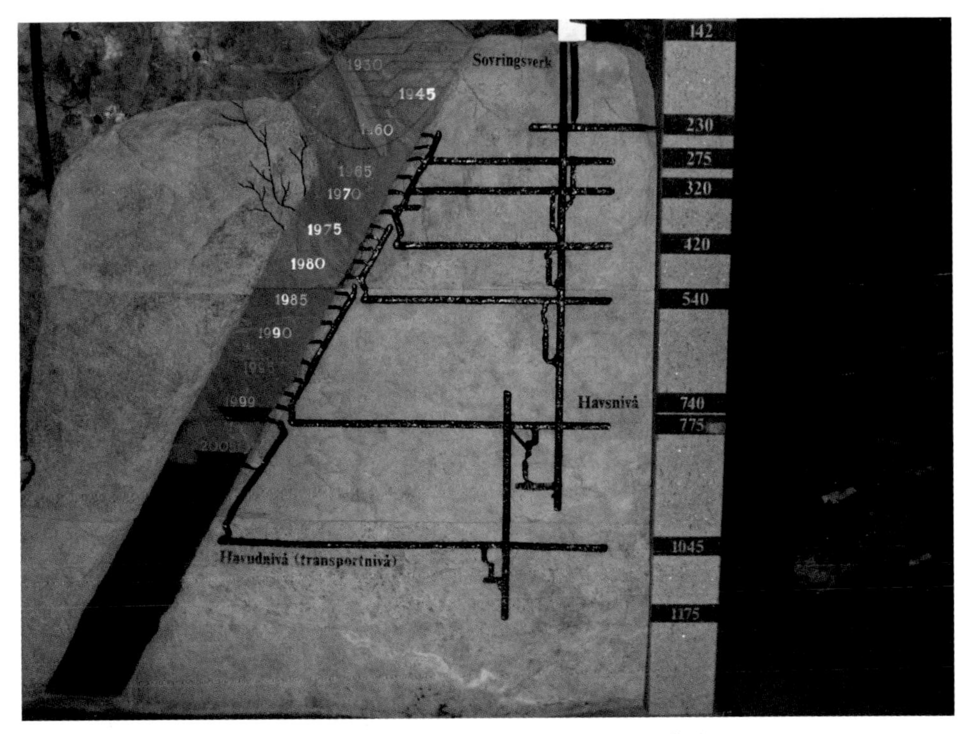

Cross-section of the Luossavaara-Kiirunavaara mine. The dark diagonal band is the vein of ore; the black lines are the working levels. The depths are given on the right.

exhausted to a depth of 700 metres. But there is still at least 1,300 metres left. Carry on mining, you might say.

But I soon change my mind when I look at the third presentation: a map of the surface, showing the city of Kiruna. It shows that the layer of iron ore passes right under the city. If the mine is excavated any further, the risk of subsidence is so great that the whole town will have to be moved. They have already started demolishing the houses closest to the mine. By 2033 the city centre will also have to be relocated. The map shows the evacuation plan. The work underground may have become a lot more pleasant but substantial sacrifices are having to be made above ground.

We get back into the bus and continue to the work face, at a depth of 700 metres. The tunnel is dark; there is only light at the work face, where the ore is drilled and blasted. There is no one to be seen, everything is automatic: that is what the people in the office are watching on their monitors. We step out of the bus in the dark tunnel and take samples of ore, magnetite, the actual iron ore, and apatite, a phosphate mineral that occurs in the ore in thin white bands. The layer of iron ore is embedded between two layers of volcanic rock almost 2 billion years old. I use my hammer to knock off two sizeable chunks. Again, the axles of the bus nearly succumb to the weight. Later, in Rovaniemi in Finnish Lapland, I send a box of stones back home, weighing 44 kilograms. I can't resist it.

The flint mines were 12 metres deep, the copper mines in the province of Hubei in China nearly 5,000 years ago were 50 metres deep, the silver mines at Laurion 100 metres, the sixteenth-century copper mines at Kitzbühel went down to 800 metres and have not gone much deeper since. Kiirunavaara is now 1,300 metres deep; they have drilled to 2,000 metres – in Falun, too – and they still haven't reached the end.

Gold mines in South Africa are nearly 4 kilometres deep, the deepest that humans have physically been underground. But it is not simple: the heat is so fierce down there that they have to pump ice and ice

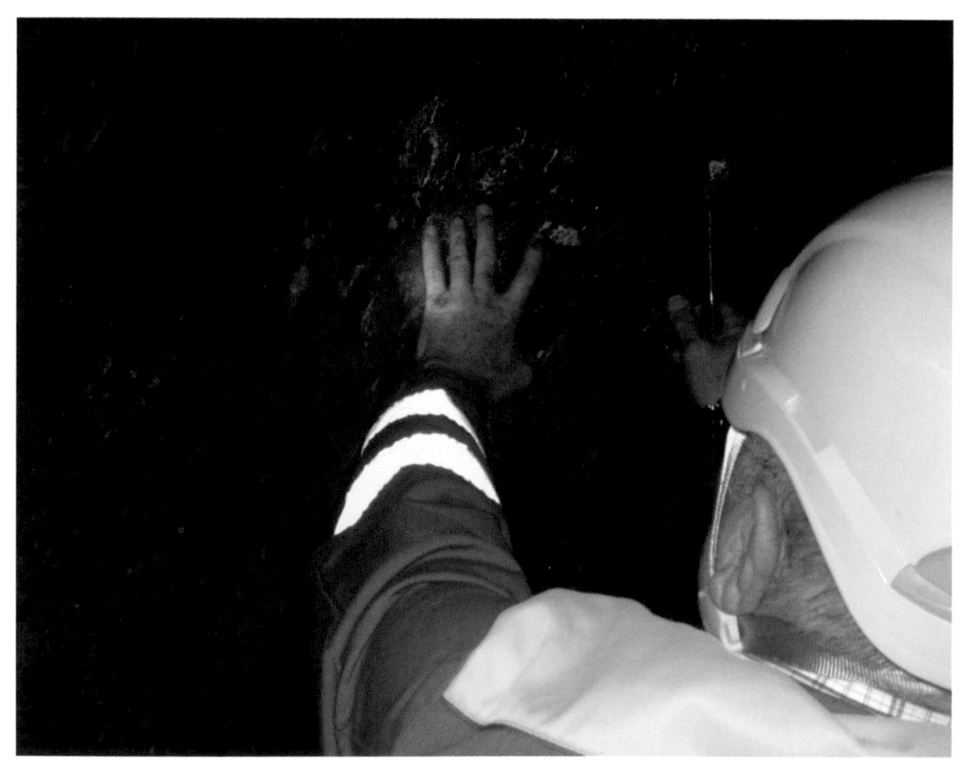

Taking samples of magnetite-apatite iron ore in the light of a headlamp, 700 m underground, Kiirunavaara mine, Sweden. The light bands are apatite, a calcium phosphate.

water into the tunnels to keep it cool, and sometimes the miners wear jackets filled with ice. Perhaps Dante wasn't so crazy, portraying Lucifer at the centre of the Earth trapped in ice. And there are plans to push on deeper to 5,000 metres.

I once promised the Mijnbouwkundige Vereeniging, the mining students' association at Delft, a bottle of vodka to the student or graduate who had been the deepest underground. The prize was won by the only entrant, mining engineer Theo Gerritsen. In Het Noorden, the mining students' own bar, they pushed a table against the wall, covered in signs stolen from mines all over the world. Theo, full of energy and built like a house, climbed up on the table and told his story.

From 1997 to 2000, he had worked as strata officer up to a depth of 2,940 metres in the South Deep mine in the South African gold region of Witwatersrand. His job was to use a compass and tiltmeter to chart cracks and fractures in the rock at the workface that might be dangerous to the miners excavating the gold ore. He told us:

> Sometimes, you have to walk for an hour through the tunnels from the vertical haft to the workface. The cooling machines near the shaft don't reach very far, so the temperature at the workface can rise to 65 or 70 degrees. The miners work stripped to the waist, wearing only long trousers, Wellington boots, a helmet and ear protectors. Some of them wear safety glasses. You have to drink large quantities of fluid all the time, otherwise you get a splitting headache from the dehydration. I drank the four litres of frozen cola I took with me in two hours. A newcomer won't stick it out for longer than 20 minutes.
>
> The worst is shaft sinking, making a new vertical shaft. They still do it with a bucket hanging on a cable, but now they are 5 metres in diameter. A team of twelve men stand in the bucket, drill a rosette of boreholes down into the rock, and fill them with dynamite. Then the bucket is pulled up a little, the dynamite goes off, they remove the rubble, descend another metre and do the same thing again. There are a lot of accidents. And that's how it goes, 800 metres vertically downwards. If you look up from the bottom, the daylight is nothing more than a white pinhole.

'*Glück auf*,' the students sang after Theo finished his story, relieved that he had lived to tell the tale. He had earned his vodka.

Anyone with a little sympathy for the deprivations of the miners, who have to hack heavy chunks of ore out of very hard rock, surrounded by darkness and fumes, bent double, squatting or lying down, will agree with Lucretius in *De rerum natura*: everything we are afraid of in the afterworld, in hell, happens here, during our lives on Earth. 'And those things sure enough, which are fabled / to be in the deep of Acheron, do all exist for us in this life' (III:978–9).

What did the early miners think about as they hacked away in the sparkling ore tunnels? Wouldn't they have wondered where all that glittering beauty came from, how it had been created? That is interesting, because you would then expect them to be curious about what is at the centre of the Earth. According to Empedocles, it would have to be a combination of the four elements: earth, air, fire and water.

Empedocles only lived 135 kilometres from Etna, and it is not inconceivable that the volcano inspired him to come up with his four-way classification: molten lava is fire, vapours are air, steam is water and solidified lava is earth, or the fuel for the fire, as suggested by Emile den Tex in his wonderful geological cultural history *Een voorspel van de moderne vulkaankunde in West-Europa*.

Aristotle worked the four elements out in greater detail. Fire is warm and dry, air is warm and wet, water is cold and wet and earth is cold and dry, meaning you can see each element as a combination of certain properties. Aristotle's pupil Theophrastus used these ideas in *On Stones* to explain the origin of 'stones' and 'earths'.

There is no denying that his views have a certain elegance. Water and cold together form ice, molten metals solidify when they are cold and melt again when they are warmed up: metals are therefore cold and wet. But salt crystals are formed by heat (the evaporation of salt water), so stones (precious stones and minerals) are dry and hot. Mud, a mixture of earth and water, can dry up through heat and cold. Earthy materials have such large pores that they can absorb water, which can be driven out again by heat.

By striking contrast to Theophrastus' concise explanation of how ore originated, in *De rerum natura* Lucretius was positively effusive.

> To proceed, copper and gold and iron were discovered and at the same time weighty silver and the substance of lead, when fire with its heat had burnt up vast forests on the great hills, either by a discharge of heaven's lightning, or else because men waging with one another a forest-war had carried fire among

Copper ore (chalcopyrite), Aitik mine, northern Sweden.

> the enemy in order to strike terror, or because drawn on by the goodness of the soil they would wish to clear rich fields and bring the country into pasture, or else to destroy wild beasts end enrich themselves with the booty; for hunting with the pitfall and with fire came into use before the practice of enclosing the lawn with toils and stirring it with dogs. Whatever the fact is, from whatever cause the heat of flame had swallowed up the forests with a frightful crackling from their very roots and had thoroughly baked the earth with fire, there would run from the boiling veins and collect into the hollows of the ground a stream of silver and gold, as well as of copper and lead. (v:1241–57)

For Lucretius, then, fire did not come from inside the earth, but was created on the surface. This is diametrically opposed to the prevailing ideas at the time about the water cycle: water was believed to come from the interior of the Earth, from the springs that can be found everywhere in the landscape. Pliny and other Latin authors added little to Aristotle and Theophrastus. Some 1,500 years later, we see the ideas of Aristotle and Theophrastus reproduced almost unchanged in *La composizione del Mondo* by Dante's contemporary Restoro d'Arezzo in 1282. The source was largely the Arab scholar Alfraganus, who also inspired Dante.

Even Agricola, a practised and accurate observer, cannot shake it off. In his other great work, *De ortu et causis subterraneorum*, also quoted extensively by the Hoovers, he attributes the origin of ore to *canales*, formed by erosion by underground waters. The voice of Plato can still be heard dimly through his text. These waters were the result of rainwater penetrating into the ground, and of the condensation of steam heated up by subterranean fires of bitumen, coal and other combustible substances. In the *canales*, water and 'juices' circulate, which are deposited in the form of 'earth', 'solidified juices', 'stone', metals and 'compounds'. 'Earth' refers to clay, soil, pigments, gypsum, marl and so on; the 'solidified juices' include salt, soda, vitriol and bitumen; 'stone' comprises precious and semi-precious stones such as quartz and amethyst; metals are those, like gold, that occur in pure form in nature; and 'compounds' are ore minerals with a metal gloss from which metal can be extracted, especially sulfur compounds like chalcopyrite and copper sulfide. Where the juices come from is not entirely clear. And so the echoes of the Greek limestone landscape still resound 2,000 years later.

It is not until we get to Descartes that the focus starts to shift again, back towards fire. In Part IV of his *Principia philosophiae*, he wrote:

> In addition to the vapours which are drawn out of waters concealed beneath the earth, many acrid spirits, oily exhalations and vapors of quicksilver ascend from the interior earth to the exterior; transporting with them particles of other metals. And all substances which are mined are formed by their mingling in various ways . . . In the same way, vapors of quicksilver, by creeping through the small cracks and fairly large pores of the earth, leave particles of other metals which were mingled with them . . . and thereby the earth is impregnated with gold, silver, lead and other metals . . . Furthermore, spirits and exhalations also transport several metals, such as copper, iron and antimony, from the interior earth to the exterior.

Descartes' gobstopper model has a specific layer from whence metals originate.

In his *Sacred Theory of the Earth* (1681), Thomas Burnet follows Lucretius more closely. In his paradisiacal, smooth, antediluvian Earth, there were no ores or minerals: no one needed them.

> As for Subterraneous things, Metals and Minerals, I believe they had none in the first Earth; and the happier they; no Gold, nor Silver, nor courser Metals. The use of these is either imaginary, or in such works, as, by the constitution of their World, they had little occasion for. And Minerals are either for Medicine, which they had no need of further than Herbs; or for Materials to certain Arts, which were not then in use, or were suppli'd by other ways. These Subterraneous things, Metals and metallick Minerals, are Factitious, not Original bodies, coeval with the Earth; but are made in process of time, after long preparations and concoctions, by the action of the Sun within the bowels of the Earth.

What is striking is that the fight between water and fire continues to this day. Since the work of James Hutton, the origin of magmas has been undisputed, but how ore is created remained a riddle for a long time. In his book *The History of the Theory of Ore Deposits* (1933) British ore specialist Thomas Crook presents a fascinating description of how the tide has ebbed and flowed between scientists over the centuries. That magmas can give off vapours was already known, but ore deposits can sometimes be so unpredictable that it is no easy task to establish the connection. Crook himself was a great proponent of the major role of atmospheric water, which may be heated up deep in the Earth and can therefore dissolve and deposit substances, but did not itself originate in the magma. But many geologists, including the American J. E. Spurr, also attributed a much greater role to magma.

Thick banks of sulfide copper ore, like those in Falun, were a particularly large problem. We became acquainted with this discussion during our excursion. It is of course not easy to know what happened 2 billion years ago, 10 kilometres deep in the crust of the Earth, but there are ways of finding out. Just as you can see from a frozen canal that the temperature must be below freezing, you can see from some minerals in rocks how high the temperature and the pressure in the Earth's crust must have been. Furthermore, there are sometimes minuscule liquid enclosures in minerals, from which you can see what liquids have passed through the rock.

My excursion guide from 17 September 1970 (I keep all these kinds of things out of a kind of nostalgia, *cuando era joven e indocumentado*, but also because I think I might need them one day – like now) records what Oen Ing Soen wrote on this subject. The Swedish geologists Geyer and Magnusson claimed that, beyond any shadow of a doubt, the copper ore at Falun must have something to do with the magma from the granite that occurs in the area around Falun. But their colleague H. J. Koark at the University of Uppsala thought it was due to exhalations on the seabed, to seawater that had somehow leached through the underlying rocks, was heated up and had absorbed copper and zinc, which it had then deposited as veins of ore between the layers of volcanic rock. From the notes in my excursion journal, it is clear that we, too, wrestled with the question.

The problem was that no one had ever seen such a thing with their own eyes. In geology, it is customary to believe the workings of a process if we can actually see it happening somewhere: ripples in rocks half a billion years old look like the ripples on a beach today, so it is probable that they were created in the same way: we call that uniformitarianism. But no one had ever seen thick banks of sulfide copper ore develop.

The solution did not present itself until the 1960s and '70s, when the deep oceans were explored and the principle of plate tectonics was discovered. In the deep sea, where the plates are moving apart along the mid-ocean ridges, 'black smokers' were discovered: plumes of hot water and hydrogen sulfide, containing high concentrations of copper, zinc and lead, rise up from the deep cracks in the ridge. The black smokers are created just as Koark had predicted, despite his not knowing of the actual existence of the phenomenon: seawater penetrates into the hot volcanic rock of the spreading zone, dissolves all the metals and deposits them as sulfides around the smoking chimney of the plume. Now we know that, we are discovering more and more similarities between old ore deposits and the underwater metal plumes. But when I was studying I had never heard of it, and I doubt whether Oen Ing Soen had read anything about it then either.

But the black smokers cannot explain all ore deposits. We still do not know, for example, whether the gold in the deep mines in South Africa was deposited as grains of gold from rivers that flowed there nearly 3 billion years ago, or found its way into the surrounding rocks later through exhalations. And yet it is important to know that, because then you can better predict how the gold-bearing rock continues deeper below the surface.

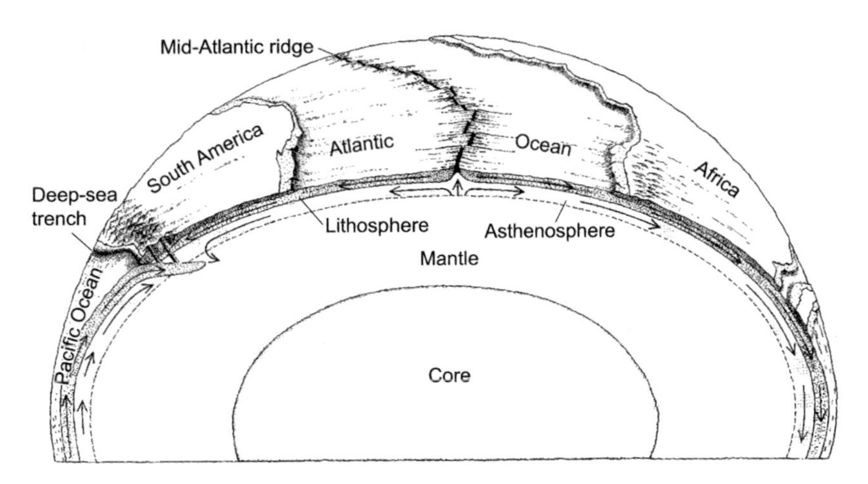

The principle of plate tectonics. A lithospheric plate is 100–150 km thick and consists of a section of the Earth's crust with the upper part of the mantle below it. Along a mid-ocean ridge, like the one between Africa and South America, the lithospheric plates are moving apart, which causes basaltic volcanism. On the west side of South America, a lithospheric plate is being forced under the South American plate (subduction), causing mountain formation, earthquakes and andesitic volcanism. The asthenosphere is the slightly viscous layer in the mantle over which the lithospheric plates move.

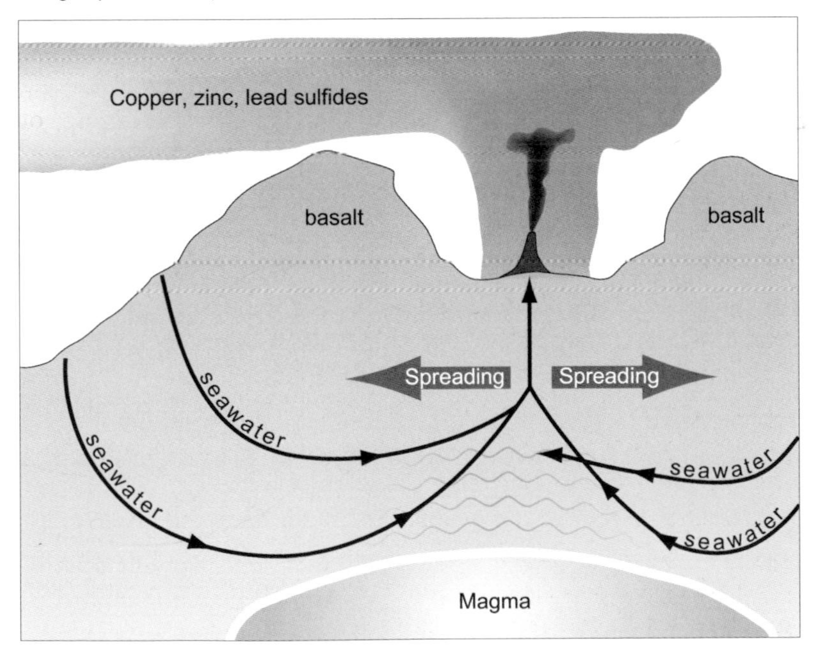

Black smoker seawater penetrates the rock along the mid-ocean ridge, dissolves the metals and sulfur compounds in it, spews them out again through chimneys in black, undersea smoke plumes, and deposits them on the seabed.

There is, however, a risk that we might never find out, because we are erasing all traces of the original location of the ore. The more efficiently we mine, the less remains as evidence of how it was originally positioned in the rock. Take the Kestel tin mine in Turkey: it is almost impossible to see that there was ever tin there, never mind discovering how it was deposited in the rock. The silver mine at Potosí is empty; all its silver is now to be found in Spanish coins and in collector's items on show in European museums. The Dutch economy was powered for more than a century by coal, but now there is nowhere where you can see a seam of coal in its natural habitat in the coalface of a mine.

Perhaps, out of respect, we should save at least one rock face with a beautiful thick seam of ore running through it, for science and for didactic reasons. Then my grandchildren can also go on a field trip to Falun, rather than having to make do with the last ever ore hand specimen in the practical room. The beauty of the underground must remain visible. But greedy mining companies do not care about that.

❧ TEN ❧

The Conflagration

And all over that sandland, a fall of slowly
raining broad flakes of fire showered steadily
(a mountain snowstorm on a windless day)
. . .
Here too a never-ending blaze descended
kindling the sand like tinder under flint-sparks,
and in this way the torment there was doubled.

Without a moment's rest the rhythmic dance
of wretched hands went on, this side, that side,
brushing away the freshly fallen flames.
Dante Alighieri, *Inferno*, XIV:28–30; 37–42

The battle between fire and water was not only how ore was created, but also about the interior of the Earth as a whole. Descartes was the first to suggest that the core of the Earth may be hot, but he also believed there was a layer of water below the Earth's crust. Just how great the confusion was can be seen from Thomas Burnet's *Sacred Theory of the Earth* (1681). Burnet thought that the Earth's current irregular surface was caused by the collapse of its crust into the underlying layer of water during the Flood. But his history does not end there. In the future, he wrote, the whole Earth will be consumed in a great conflagration, partly from the Sun but also from the Earth's internal fire. Burnet does not explain where this internal fire comes from. At one point, he compares the Earth with an egg, with the yolk replaced by a kind of central fire, adding 'which, though very reasonable, we had no occasion to take notice of in our Theory of the Chaos'.

Burnet also finds it rather difficult to explain how the entire Earth can be consumed by fire. In his defence, he does restrict himself to purely rational arguments:

The difficulty, no doubt, will be chiefly from the great quantity of water that is about our Globe; whereby Nature seems to have made provision against any invasion by fire, and secur'd us from that enemy more than any other. We see half of the Surface of the Earth cover'd with the Seas: whose Chanel is of a vast depth and capacity. Besides innumerable Rivers, great and small, that water the face of the dry Land, and drench it with perpetual moisture. Then within the bowels of the Earth, there are Store-houses of subterraneous waters: which are as a reserve, in case the Ocean and the Rivers should be overcome. Neither is water our onely security, for the hard Rocks and stony Mountains, which no fire can bite upon, are set in long ranges upon the Continents and Islands: and must needs give a stop to the progress of that furious Enemy, in case he should attack us. Lastly, the Earth it self is not combustible in all its parts. 'Tis not every Soyl that is fit fewel for the fire. Clay, and Mire, and such like Soyles will rather choak and stifle it, than help it on its way. By these means one would think the Body of the Earth secur'd; And tho' there may be partial fires, or inundations of fire, here and there, in particular regions, yet there cannot be an universal fire throughout the Earth. At least one would hope for a safe retreat towards the Poles, where there is nothing but Snow, and Ice, and bitter cold. These regions sure are in no danger to be burnt, whatsoever becomes of the other climates of the Earth.

As to the Central Fire, I am very well satisfied it is no imaginary thing . . . And tho' I do not know any particular observation, that does directly prove or demonstrate that there is such a mass of fire in the middle of the Earth; yet the best accounts we have of the generation of a Planet, do suppose it; and 'tis agreeable to the whole Oeconomy of Nature; as a fire in the heart, which gives life to her motions and productions . . . This Central Fire must be enclos'd in a shell of great strength and firmness; for being of it self the lightest and most active of all Bodies, it would not be detained in that lowest prison without a strong guard upon it. 'Tis true, we can make no certain judgment of what thickness this shell is, but if we suppose this fire to have a twentieth part of the semidiameter of the Earth . . . there would still remain nineteen parts, for our safeguard and security. And these nineteen parts of the semidiameter

> of the Earth will make 3268 miles, for a partition-wall betwixt us and this Central Fire.

Burnet writes that there are no signs that the internal fire is attempting to escape its prison deep in the Earth. But the world's volcanoes are only a prelude to the great conflagration, designed intentionally by Providence to keep us alert as a constant warning of what will eventually come to pass. About the terrible flows of fire during the eruption of Etna in 1669, about which he read in the works of the mathematician Alphonsus Borellus of Pisa, he says:

> This is beyond all the infernal Lakes and Rivers, Acheron, Phlegeton, Cocytus, all that the Poets have talkt of. Their greatest fictions about Hell have not come up to the reality of one of our burning Mountains upon Earth.

Burnet urges us to imagine that all the Earth's volcanoes raging at once and all 'lakes of pitch and brimstone and oily Liquors' bursting into flame and fuelling the fire.

> So we must expect new Eruptions, and also new sulphureous Lakes and Fountains of Oyl, to boyl out of the ground; And these all united with that Fewel that naturally grows upon the Surface of the Earth, will be sufficient to give the first onset, and to lay vast all the habitable World, and the Furniture of it.

After the whole Earth has been consumed by the conflagration, it will once again be completely smooth and resemble its original paradisiacal state. According to Stephen J. Gould, Burnet's history of the Earth is not only a linear progression in time, Time's Arrow, but also cyclic, Time's Cycle.

In his new Paradise only the just have survived. The souls of the sinners have been devoured in the conflagration. Burnet does ask, however, how the souls of those already dead can be burned: a scientist's question. And of course, after the Apocalypse, there would no longer be any need for hell, as all sinners would already have been burned. As British historian Daniel P. Walker shows in his book *The Decline of Hell*, whether hell was temporary or eternal was an issue that not only Burnet but also theologians and the whole Church had wrestled with since the dawn of Christendom. If it is eternal, it is also pointless, since showing remorse doesn't help one bit and you think only of the burns you suffer and not of your evil doings.

All of Descartes' and Burnet's visionary ideas about the Earth are ultimately based solely on scholarly knowledge and thought. Neither went out into the field to make their own observations. This cannot be said of Athanasius Kircher, whom we encountered previously at the Grotta del Cane.

Kircher was an intriguing person, a German Jesuit who spent a large part of his life living at the papal court in Rome, and who wrote several dozen books on widely varying subjects. Each of the books themselves covered a broad range of topics, including music, archaeology, language, mathematics, astronomy, optics, magnetism, geology, cabbalism, alchemy and cryptography. And all to the glory of God, he writes, but actually more to his own glory. He felt that everything was connected to everything else; he was a holist *avant la lettre*.

He claimed to be have discovered the plague bacillus at a time when microscopes were not yet good enough to be able to see it. He claimed he could decipher Egyptian hieroglyphs, translating seven signs on the Pamphili Obelisk in Rome as: 'The creator of all vegetation and fruitfulness is Osiris, whose generative force holy Mophta draws into his kingdom from heaven.' This made him world famous. What the inscription really says became clear two centuries later, after Champollion had deciphered Egyptian script using the Rosetta Stone: the signs stand for 'Autokrator', a Greek name. This is all described in Anton Haakman's hilarious book, published in 1991, on Kircher and two of his modern, equally dubious admirers. In 1974 Haakman made a wonderful documentary based on the book, which was last shown on television in 1991.

Kircher designed countless devices, not all of which excelled in their usefulness. One was a 'cat piano', a series of small compartments each containing a cat with its tail sticking out. It was played by standing on the cats' tails. He also claimed to be the spiritual father of many inventions by others. His books are full of home experiments intended to support his claims. With Kircher, science, fantasy, plagiarism and deceit are all tangled up together.

Anton Haakman concludes that it was already clear in the eighteenth century that Kircher's theories were based on hot air. He was even the butt of criticism from most of his contemporaries. As Paula Findlen recounts in her biography *Athanasius Kircher: The Last Man Who Knew Everything*, fellow Jesuit and mathematician Antonio

Baldigiani said that he 'no longer discussed matters of science with Kircher' because he was 'afraid of seeing [him]self published one day in one of his books as the author and witness of some gross error'. Dutch mathematician and inventor Christiaan Huygens called Kircher's *Iter extaticum*, which described a journey into space and to the subterranean world, 'nothing more than an accumulation of futile humbug and sheer nonsense'. The famous mathematician Leibniz, another contemporary, remarked: 'He understands nothing.' After reading parts of *Mundus Subterraneus*, Henry Oldenburg wrote in a letter to Spinoza: 'All his arguments and theories are no credit to his wit.' Newton did not condescend to quote Kircher even once. Kircher's library and museum have disappeared, though, according to Eric Asselborn, curator at the École des Mines in Paris, there are still two handpieces from his collection in the museum at La Sapienza University in Rome: a piece of solid silver and a chunk of deep-red, sparkling proustite, a silver ore, both from South America.

At the end of his book, Anton Haakman quotes one of Kircher's admirers: 'I now believe that [Kircher] was so captivated by his research that he could no longer distinguish truth from falsehood. That, in the heat of his investigations, he himself believed what he wrote, because he wrote it in a state of ecstatic inspiration.' Haakman chooses his quotation with his tongue in his cheek, as it applies equally to the admirer himself. And Paula Findlen said in 2004: 'He was a man unable to recognize truth from falsehood, a scholar with an imperfect grasp of the science of philology and linguistics, an archaeologist who did not know the difference between a Roman lamp and a Grecian urn, and an inventor of language who could not recognize the simplest cipher.'

And yet – and this is what is fascinating about the man – he was, in a certain sense, prescient. He had a visionary notion about some things which later proved to go in the direction he had suspected. He was, in any case, the first who did *not* assume that there was a layer of water below the Earth's crust, and did base his theories on the all-embracing presence of subterranean fire. His engraving of the interior of the Earth has become famous, and it is striking how closely it resembles modern views on what the Earth is like below our feet, as we shall see later.

> That there are Subterraneous Fires no sober Philosopher can deny; if he do but consider the prodigious Vulcano's, or fire-belching Mountains; the eruptions of sulphurous fires not

Kircher's view of the subterranean world, showing the fire canals fed by the central fire, which force their way up through cracks in the rock to the Earth's surface, where they erupt outwards as volcanoes. From his *Mundus Subterraneus* (1664).

only out of the Earth, but also out of the very Sea; the multitude and variety of hot Baths everywhere occurring. And that they have their source and birth-place, not in the Air, not in the Water; nay, nor as the Vulgar persuade themselves, not at the bottom of the Mountains; but in the very in-most privy Chambers, and retiring places of the Earth, is as reasonable to think . . . For how else could there be every such a quantity of Minerals, brimstone, and sulphurous unctuous matters, without any fire and subterraneous burnings of fire-engendring, and all concocting nature; which by no means can be conceived to be enkindled, from the conflicts of air and moisture, in those most dark and deep Regions of the Earth, so remote from all influences of the Sun?

This passage is typical of his self-satisfied prose, which he uses to wipe the floor with anyone who does not agree with him in decidedly unscientific terms, just like Karl Ernst von Baer much later. He claims, for example, that Democritus believes the Earth's heat to come from limestone, and from 'antiperistasis', the process by which hot and cold reinforce each other. Kircher finds this such a foolish theory that it has to be disputed 'not by reason, but by allowing it to be jeered and laughed at by old women'.

It is nevertheless interesting to investigate where Kircher's ideas about the 'subterraneous fire' come from, which he explains in the preface to *Mundus Subterraneus*. In 1638 he saw Etna and Stromboli erupt from on board a ship ('I saw that Mount Etna and the island of Stromboli expelled extraordinarily large pieces of rock and smoke, as large as whole Mountains'), and shortly afterwards he climbed Vesuvius at night, less than seven years after the great eruption of 1631:

> I had a great desire to know whether Vesuvius also had not some secret commerce and correspondence with Strongylus [Stromboli] and Ætna . . . When I had arrived, I saw it all over of a light fire, with an horrible combustion, and stench of Sulphur and burning Bitumen. Here forthwith being astonished at the unusual sight of the thing; Methoughts I beheld the habitation of Hell; wherein nothing else seemed to be much wanting, besides the horrible fantasms and apparitions of Devils. There were perceived horrible bellowings and roarings of the Mountain; an unexpressible stink; Smoaks mixt with darkish globes of Fires; which both the bottom and sides of the Mountain continually belch'd forth out of Eleven several places; and made me in like manner, ever and anon, belch, and as it were, vomit back again, at it: O, the depth of the Riches of the Wisdom and Knowledge of God! How incomprehensible are thy wayes! If thou shewest thy power against the wickedness of mankind in so formidable and portentous Prodigies and Omens of Nature; What shall it be in that last day, wherein the Earth shall be drown'd with the Ire of thy Fury and the Elements melt with fervent heat?

In the morning Kircher looked down from the edge of the crater and saw on its floor the new cone created during the eruption of 1631, and

Vesuvius as Kircher saw it with his 'inner eye', says Dieter Richter somewhat derisively in his book *Der Vesuv Geschichte eines Berges*. The open section at the front is designed to show that the fire really does come from the interior of the Earth.

the 'Earth, Sulphur and other Minerals [fell] to the ground, arousing so great and fearful a sound, as that any, even of the stoutest and most undaunted heart, would scarce venture to suffer'.

Conflagration is the punishment imposed on the giant Capaneus in the third ring of the seventh circle of Dante's Inferno. But why? First we hear from Capaneus himself, and then from Virgil, as Dante's guide through hell.

> 'Let Jupiter wear out his smith, from whom
> he seized in anger that sharp thunderbolt
> he hurled, to strike me down, my final day,
>
> let him wear out those others, one by one,
> who work in the soot-black forge of Mongibello
> (as he shouts, "Help me, good Vulcan, I need your help,"

the way he cried that time in Phlegra's battle),
and with all his force let him hurl his bolts at me,
no joy of satisfaction would I give him!'

My guide spoke back at him with cutting force,
(I never heard his voice so strong before):
'O Capaneus, since your blustering pride

will not be stilled, you are made to suffer more:
no torment other than your rage itself
could punish your gnawing pride more perfectly.' (XIV:52–64)

Pride, yes, that is what Kircher suffered from the most. But there is one thing you do have to allow Kircher and his contemporaries: none of them suggested that what lies beneath the ground really is hell. Hell is an easy metaphor for volcanism and a warning of the fire that will consume the world on the Day of Judgment, but it is not an underground realm of the dead, where you will go to if you are not careful, and where past sinners are already being punished for their bad deeds. In the seventeenth century hell gradually declined in importance, not only because mercy gradually became more important than retribution, as Walker argues in *The Decline of Hell*, but also through the tumultuous development of science. But the Jesuit Athanasius ('the immortal') Kircher, in the pay of the pope, was careful to remain very superficial about the real hell. Nevertheless the Catholic Church refused to canonize him: he had come too close to hell when he climbed Vesuvius.

ELEVEN

The Monster Geryon

And now, behold the beast with pointed tail
that passes mountains, annulling walls and weapons,
behold the one that makes the whole world stink!

These were the words I heard my master say
as he signaled for the beast to come ashore,
up close to where the rocky levee ends.

And that repulsive spectacle of fraud
floated close, maneuvering head and chest
on to the shore, but his tail he let hang free.

His face was the face of any honest man,
it shone with such a look of benediction;
and all the rest of him was serpentine;

his two clawed paws were hairy to the armpits,
his back and all his belly and both flanks
were painted arabesques and curlicues.

Dante Alighieri, *Inferno*, XVII:1–15

At the end of the seventeenth century the idea began to take hold that the Earth may be hollow. That was caused by an error made by Isaac Newton in the first edition of his major work of 1687, *Philosophiae naturalis principia mathematica*, often referred to simply as the *Principia*.

Copernicus had deduced that the Earth rotates around the Sun; Galileo had defied the Inquisition to support that argument and designed a revolutionary telescope with which he discovered the moons of Jupiter; Kepler had demonstrated the elliptic nature of the orbits of the Earth and the other planets; but Newton was the first to understand the driving force behind all these phenomena: gravity. It was the most important scientific discovery of the Enlightenment.

This all passed by the Jesuit Athanasius Kircher. He remained faithful to the geocentric world view of the Catholic Church, even though he must have been familiar with Galileo's famous books *Dialogo sopra i due massimi sistemi del mondo* from 1632 and *Discorsi e dimostrazioni matematiche intorno à due nuove scienze* from 1638, and surely knew of his conviction by the Inquisition in 1632, because he was working for the Papal Collegium Romanum in Rome at the time.

Newton's law of gravity stated that two bodies attract each other with a force equal to the product of both their masses, divided by the square of the distance between them, multiplied by a constant, known as the gravitational constant G. This made it possible to calculate the masses of celestial bodies like the Earth, the Moon and the Sun.

Newton worked out the gravitational force between the Earth and the Moon on the basis of the tides, the twice daily rise and fall being caused by the Moon and, to a lesser extent, the Sun pulling the waters of the Earth a little towards them. He took as his starting point the difference between the spring tide and the neap tide at Plymouth and in the Bristol Channel, and calculated that the Earth's mass was 26 times greater than that of the Moon, while the Earth's density was 5/9 that of the Moon's. That was a surprising result, since it would mean that the Moon's rocks were, on average, twice as heavy as those of the Earth.

Newton's contemporary, Edmond Halley – famous for the comet that bears his name – thought that strange and wrote an article in 1692 stating that the only explanation for the difference in density was that the Earth must be hollow. The English scientific historian Nick Kollerstrom suggests that this idea might have its origins in Thomas Burnet's water cavities in the Earth. It later emerged that Newton had miscalculated: the mass of the Earth is in reality 81 times greater than that of the Moon, and its density is 5.9 compared to 3.3 grams per cubic metre. By then, however, the hollow Earth theory had achieved widespread popularity.

Halley believed that there was a second sphere within the Earth which rotated independently of the outer shell that is known to us. This, too, was a visionary idea, as we shall see. The outer shell was 500 miles thick, and there was a space of another 500 miles before you came to the inner sphere. Perhaps there were even a series of spheres. Not everyone agreed with the theory, suggesting that the oceans would leak down to the interior if cracks appeared after an earthquake. Halley disputed this, saying that the water was held in place by 'saline and

Michael Dahl, portrait of *Edmond Halley* at 80 years of age, with a diagram of his hollow Earth, 1736.

Vitriolick Particles as may contribute to petrefaction'. He also thought that the subterranean space was habitable. The theory was eagerly taken up by science fiction authors and crackpots like John Cleves Symmes, who claimed in 1818 that there is a large hole at the poles where you can enter the inner world. These ideas survive to the present day. Edgar Rice Burroughs, creator of Tarzan, enjoyed great success with his books on the inner planet Pellucidar, including *At the Earth's Core* (1914), which was made into a film in 1976.

One of the most enjoyable books about the inhabitants of the hollow Earth is *Icosaméron*, written in 1788 by Giacomo Casanova, the legendary seducer. Casanova was also a gifted violinist, diplomat, spy, gambler, actor and writer. He wrote the book in French. Its title means 'twenty days' because the story is told in 20 days, just as Boccaccio's *Decameron* is told in ten.

In 1533 brother and sister Edouard and Elisabeth sail from Plymouth on the ship *Wolsey* to accompany their uncle on a voyage to explore the northern seas, discover new lands and find a northern passage. From a distance they see the volcano Hekla erupting on Iceland. 'On that island the people believe that this volcano is the mouth of

John Cleves Symmes's hollow Earth.

hell, and are certain of it, as they have seen with their own eyes how bands of devils often enter there with doomed souls in their claws, to emerge again later to seek others.' (Iceland is also the place from where, a century later, Jules Verne's heroes Professor Lidenbrock and his nephew Axel begin their journey to the centre of the Earth.)

The *Wolsey* sails on and gets caught in the Maelström, the powerful current near the Lofoten Islands off the Norwegian coast. 'Maelström' is a French-Norwegian corruption of an originally Dutch word meaning 'whirling stream'. It is the legendary maelstrom that Kircher also describes in *Mundus Subterraneus*, which he thought was the entrance to a subterranean canal linking the North Sea to the Baltic and the Barents Sea. It was a notorious spot among seafarers and, according to legend, many ships had been sucked down to their doom by it. Modern Norwegian marine research has shown that there are eddies caused by the strong tidal currents passing over a shallow fjord sill, but they are not as dramatic as they are portrayed to be in the old literature.

In his renowned story 'A Descent into the Maelström' (1841), Edgar Allan Poe describes how a fishing boat with three brothers founders there. The youngest brother has lashed himself to the mast and disappears into the depths when it breaks off. The second holds tight to a ring-bolt, closes his eyes and is also sucked down. But the third brother, sitting on a cask, studies the rotation of the whirlpool closely and, by following the slowest of the downward movements exactly, manages to escape. In his inspiring book *Terre*, Peter Westbroek uses the behaviour of the second and third brothers as a metaphor of how we treat the Earth. If we blindly continue to act recklessly, it will go wrong, but if we are careful we will be able to survive.

Edouard and Elisabeth also take great care when the *Wolsey* finds itself in the whirlpool. On the deck they find a lead chest, the property of an old seaman who, if he dies at sea, wants to be placed in the chest and thrown overboard so that the sea monsters will not be able to tear him apart. He plans to await the Last Judgment in safety on the seabed. The chest is large enough for twelve bodies and has portholes on all sides so that you can look outside. It is open, so they look inside; the inner walls are padded, and it contains bottles of water and brandy, an atlas, a compass, a Latin Bible, two pairs of pistols, gunpowder, bullets, a pointer case, an operating set, Indian ink and a paintbox with pastel colours, and tongs of different sizes. In a corner is the largest lodestone they have ever seen. The ropes holding the chest to the deck suddenly break and, whipping wildly, strike Edouard and Elisabeth so that they fall into the chest, and it slams shut. It is then carried away into the depths of the maelstrom.

Instead of ending up on the seabed, they keep falling until a fiery light suddenly shines through the portholes, and it becomes boiling hot. The chest continues to fall, and it becomes cold and wet again. Then it is hot once more, and the smell of sulfur penetrates the chest. It is struck by something and tips over, they become weightless, and the large lodestone floats to the ceiling of the chest. Suddenly it falls back to the floor, but they themselves are upside down. Gravity is now working in the other direction. The chest has come to a halt in the hollow interior of the Earth's crust. Red asexual beings, no taller than a yard, with white blood and a single green nipple on their breasts, save them from the red water in which they have landed. Edouard and Elisabeth regain their strength by drinking the red milk from their saviours' nipples.

Welcome to the small inhabitants of inner earth, the Mégamicres, seemingly given this name by Casanova to contrast them with the extraterrestrial giant Micromégas, who visits Earth in the novel of the same name by his arch-enemy Voltaire (1752). In the land of the

Mégamicres the sun shines all the time; it hangs directly above their heads and there is no day or night. It is the magnetic iron core of the Earth for, as soon as they have left the chest, it shoots lodestone and all towards the glowing inner sun. We are back in Virgil's underground Elysian Fields, which also had their own sun. According to Edouard, the diameter of the inner Earth is 188 miles smaller than ours, meaning that the Earth's crust is 175 kilometres thick; not as thick as Halley said, but a good estimate, as we shall see. It is obviously interesting to let yourself be swept away by a maelstrom.

Edouard and his sister stay under the ground for 81 years. They learn to understand the complex yet harmonious society of the Mégamicres, witness their political developments, have children together and, because reproduction takes place much more quickly there, they have a total of 4 million descendants. But they do not get older; they retain their eternal youth until they return. It is not hell that can be found in the interior of the Earth, they say, but Paradise.

One day Edouard wants to follow a gold lode in an underground mine for a friend who is a Mégamicre senator, so he prepares a charge of *poudre verte et bleue* to blow up the rocks. But instead of a fountain of stones flying through the air, such a large area of the surrounding land goes up in one go that even the piece on which he and Elisabeth, who has hurried to his side, are standing is thrown into the air. When they come down again, they fall into a pitch-black cavern under the ground, together with a brightly variegated Mégamicre. When they try to get out of the cave, they discover that they are in a tunnel that slowly rises towards the surface of their own Earth. By the light of their carbuncles they follow the tunnel for many weeks, plagued by hunger, thirst and bleeding feet. But then they find a subterranean lake, and then another, and follow an underground river. The trusty Mégamicre dives for shellfish and crayfish for them to eat. The sand on the shore of the lakes proves to contain oil, so they can also cook the fish. Sometimes they find fennel and lilies and eat the bulbs. They experience subterranean volcanic eruptions and an earthquake, and finally they reach the surface.

Suddenly, among the French text of the *Icosaméron*, are the following words: *E quindi uscimmo a riveder le stelle*: and we came out to see once more the stars. I recognize it with a shock: it is the last line of Dante's *Inferno*. The second shock follows immediately: they emerge near the village of Zirchnitz in the region of La Basse Carniole. Lake Zirchnitz, now Cerknica in the south of Slovenia, is an interesting lake: from the autumn to the spring, it replenishes itself with winter rains and smelt water from the mountains, while in the summer it is dry and crops grow on its bed. It is a polje in the middle of the classic karst mountains, close to the world-famous caves of Postojna, in limestone from the Jurassic and Cretaceous eras. The subterranean lakes and rivers of the *Icosaméron* therefore really exist, just like the Maelström itself. Athanasius Kircher also writes about Lake Zirchnitz in *Mundus Subterraneus*: in a section headed 'From the wondrous world of Carniolia', he describes how the water runs away 'through fine discharges through a subterranean opening' – a ponor! Casanova, born in Venice and familiar with Trieste, undoubtedly knew those caves and the Lake, or had at least read Kircher. He knew his classics. It's almost enough to make you believe in the Mégamicres.

Even the shellfish and the crayfish in the subterranean rivers and lakes are not necessarily the products of Casanova's imagination. The caves at Postojna are a hotspot for subterranean biodiversity in Europe. Biologists call the creatures that live underground, often in caves and groundwater, 'stygobionts', the biota of the Styx. Stygobites live purely under the ground; stygophile fauna, partly above and partly below the ground; and those that only go underground occasionally to seek protection are called stygoxenes. Recent surveys of European karst caves have counted 1,239 species that live entirely underground, mainly crustaceans and crayfish-type creatures. It is as though Casanova already knew.

Someone else who knew was my uncle, Lipke Bijdeley Holthuis. He had collected exotic shrimps in caves all over the world. As a boy, I always wanted to go with him, but was never allowed to. In his workroom the shrimps hung motionless in their cylindrical jars, preserved in alcohol until it was their turn under the microscope. And then they

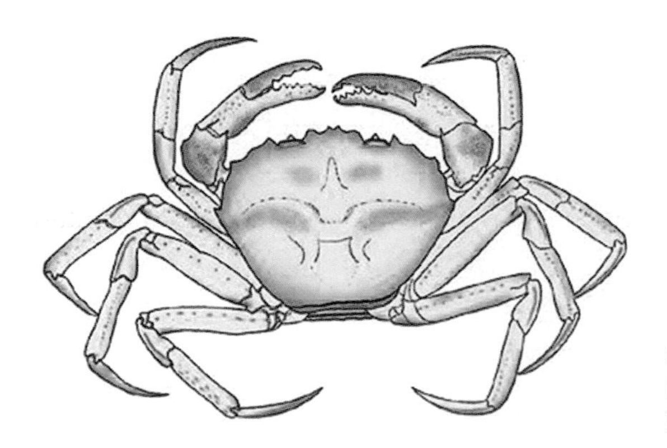

Geryon fenneri (Manning and Holthuis, 1984).

Gustave Doré, *Geryon*, 1857.

were meticulously drawn, down to the last minute detail, and prepared for eternity. I didn't see those shrimps again until he died in 2008: in his publications and on the Internet, about the crayfish, crabs, shrimps, woodlice and water fleas of Ambon, Mexico, Morocco, the United States, Israel, Thailand, Borneo, New Guinea, the Netherlands Antilles, Japan, Suriname, Tristan da Cunha and the Bahamas. He discovered dozens of new species, all of which – as the rules prescribe – were given the third name Holthuis, such as *Edoneus atheatus* Holthuis, a shrimp from a cave in the Philippines. Together with his colleague Raymond B. Manning from the Smithsonian Institution, he described new deep-water crabs from the Bahamas of the species

Geryon: named after the monster that flew Dante and Virgil across the gorge from the seventh to the eighth circle.

A total of 60 creatures have been named directly for him, having *holthuisi* as their second name. He didn't do this himself, because it is not done to name an animal after yourself: this is something you let others do. There is even a shrimp from New Zealand that has been named *Lipkius holthuisi*. But my favourite is *Stygiomysis holthuisi* Gordon, a kind of mysid that Lipke had collected in a cave on Sint Maarten, which he had sent to Isabella Gordon at the British Museum – a woman who wore long brown dresses, if I remember correctly – to be identified. Isabella had initially called the creature *Rhopalonurus holthuisi*, but later renamed him Holthuis's mysid from the Styx. *Stygiomysis holthuisi* is the only real inhabitant of the Styx in the family.

Stygiomysis holthuisi (Gordon, 1958).

Uncle Lipke wrote more than 600 articles about his crustaceans, a massive volume of work for a scientist, even though many of them were about incorrect names that had been given to creatures in the past. And, as he confessed to me shortly before he died, he had never given a presentation at a scientific congress. He didn't like talking and preferred to write everything down. Nevertheless, he was the world's leading expert on crustaceans.

Although he was not a fan of the theory of evolution, his colleagues were more broad-minded about it. I found an article by the same Manning from the Smithsonian, who proposed that the famous Stygiomysis mysids represented an ancient genus from the Jurassic or Cretaceous periods – before America and Europe drifted apart, because they can be found on both sides of the Atlantic Ocean. Yet another missed opportunity for Lipke and I to come together; a pity that I didn't know about it before.

Incidentally, you do not need caves to encounter subterranean life: there are enough in the soil, the groundwater, and the pores and cracks in rocks, down to great depths. The soil remains one of the least known ecosystems around the world. According to a recently published book, *De Nederlandse biodiversiteit*, there are 86 different kinds of woodlice in the Netherlands alone. The book summarizes its content as follows: the biodiversity of the Netherlands consists of worms and insects. There are around 47,800 species of organisms in the Netherlands, and it is thought that a further 10,000 have yet to be discovered.

But the number of species is not in itself particularly interesting. The main question is: what do all those species live on? What niches have all those 86 kinds of woodlice found for themselves in the soil? Or are they permanently at loggerheads because they all have their sights set on the same worm droppings? We know quite a lot about what rabbits, deer and wild boars do in our ecosystems, but we know precious little of what all those worms, springtails, nematodes, mites and woodlice do in the ground. And there is very little chance that we will ever be able to study them in their natural environment, because in the Netherlands there is not a single square metre of ground that has not been ploughed, harrowed, drained, fertilized, planted or excavated for agriculture, urban development, hydraulic engineering works, conservation projects or any other number of activities that humans have devised to destroy the soil.

And anyone who thinks that there is only life in the upper layer of the Earth's surface is gravely mistaken. Caves can go as deep as 2 kilometres underground, and they all contain life: blind fish, blind lizards, blind frogs and a whole spectrum of small fry that no one has yet set eyes on. Bacteria being transported downwards with leaching groundwater a metre a year below the surface have time on their side: in 1,000 years they can reach a depth of 1 kilometre. There may be little food down there, but bacteria do not need much: they can usually survive without oxygen, and live by reducing iron oxides or sulfate from the rock. In a drill-hole in Sweden researchers have discovered heat-seeking bacteria at a depth of more than 3 kilometres, at temperatures from 65–75°C. In the same South African gold mines we discussed earlier, sulfate-reducing bacteria live at depths of 2.7 to 3.4 kilometres under the ground, sometimes in extremely alkaline conditions. During deep-sea drilling, bacteria have even been discovered a kilometre below the seabed. And they seem to play a role in the generation of methane from oil in deep reservoirs. So it's actually quite busy down there, under our feet. Researchers are currently devoting a great deal of attention to these extremophiles, as they may help us to search for

Burrows dug in weathered granite by termites, Kenya.

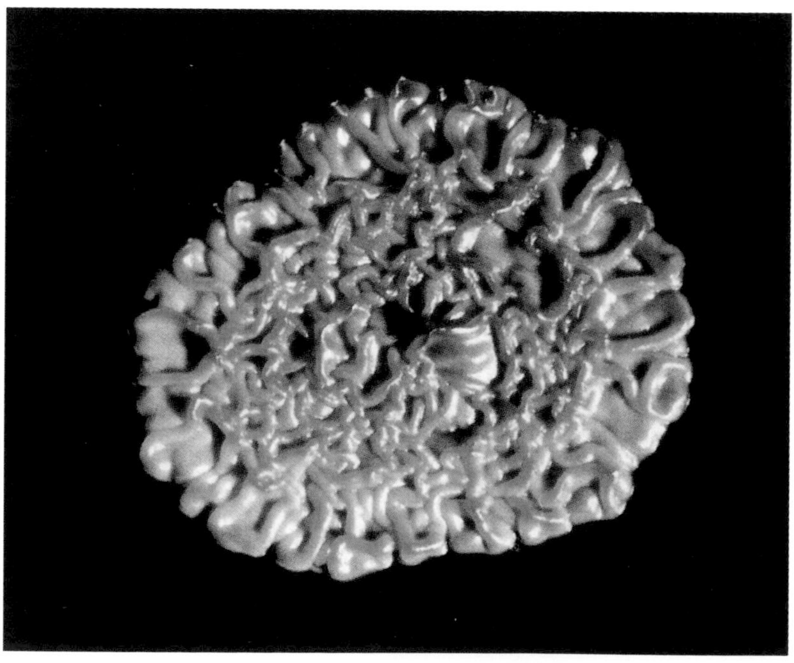

Colony of bacteria isolated from unsaturated sediment at a depth of 122 m, National Engineering Laboratory, Idaho.

extraterrestrial life. But it also means that our terrestrial ecosystems go a lot deeper than we first thought, and we need to take them into account in large-scale subterranean interventions. And we haven't even yet been down as deep as the Mégamicres lived: 175 kilometres.

TWELVE

The River of Tar

Here, too, but heated by God's art, not fire,
a sticky tar was boiling in the ditch
that smeared the banks with viscous residue

I saw it there, but I saw nothing in it
except the rising of the boiling bubbles
breathing in air to burst and sink again.
Dante Alighieri, *Inferno*, XXI:16–21

After much searching I find the institute. It is hidden behind a petrol station and small shops selling building materials. The high, rusty gate is hanging at an angle. The gateman is an old man of at least 80. He doesn't want to let me in. He speaks no Russian, only Azeri. I show him the visiting card of the director, Fuad Mahmudoglu Haji-Zadeh. The old man rummages around in a drawer, pulls out a sheet of paper from a school exercise book with numbers written on it all higgledy-piggledy, and calls a three-figure number that is not on the business card. Then I am allowed to enter.

Nothing in the dilapidated inner courtyard suggests that this is the Azerbaijan National Aerospace Agency. But that is the intention: in Soviet times, it was secret. Now that its lifeline to Moscow has been cut off, the institute is in dire straits. The director lets me in immediately and offers me tea with bonbons the size of dominoes. I tell him I need aerial photos. No problem, he says, and looks at a handwritten list. Then he searches through a large pile of cardboard files on another desk and produces a few of the photos I need. The second series is more difficult. A woman comes in with a thick roll of negatives. She hands me one end of the roll and, as though we are folding up a table-cloth together, we roll the negatives out and hold them up to the light of the window to see which ones I want. I'd like these ones, I tell the director. No problem, he says, that will cost you $300. And I have to pay in advance, otherwise he cannot buy the paper to print them on. He tells me they will be ready the day after tomorrow. And they were, too.

That was in 1999. Now I am at my computer in Delft and I can retrieve the same pictures from Google Earth within a few seconds. Perhaps they are not as sharp, but I see the same thing as I did then: the old oil wells of Baku. Not boreholes, but real wells that have been dug. Before oil was extracted by drilling, it used to be excavated by digging shafts.

The Sumerians in Mesopotamia already knew about oil or tar and used it as cement to hold building bricks together, to burn in lamps, as a lubricant, as a medicinal salve and to make gutters and pipes waterproof. The builders of the Tower of Babel said to each other: 'Go to, let us make brick, and burn them thoroughly. And they had brick for stone, and slime [tar] they had for mortar' (Gen. 11:3–4). At first they didn't even need to dig or drill: the oil was easy to find on the surface, in an oil seep. Marco Polo wrote about Azerbaijan: 'there is a spring from which gushes a stream of oil, in such abundance that a hundred ships might load there at once. This oil is not good to eat, but it is good for burning.'

Sometimes natural gas, too, just comes out of the ground, as Alexander the Great had observed. English merchant and discoverer Jonas Hanway described in 1746 how the inhabitants of the Absheron peninsula near Baku each had a gas pipe in their huts closed off with a clay plug. When they wanted to cook, they removed the plug, ignited the gas and warmed up the food. When they were finished, they replaced the plug until the next time. Everyone had, as it were, their own gas field under their house.

Oil was in great demand and, for many centuries, there was a lively trade in it that extended to Persia, Central Asia and Turkey. The Azeris dug wells to extract oil as early as the sixteenth century. They would make a wooden shaft sometimes 15 metres deep and lower a man into it with a leather water bag which he had to fill with oil, in the same way in which the Uyghurs in China dug for coal. He was pulled back up to the surface using men or horses. This could be dangerous, as the oil was often under pressure, and many miners drowned because the oil at the bottom of the well suddenly began to boil, like Dante's corrupt sinners splashing around in the river of tar – except that here it was more likely to be the bosses that were corrupt than the poor wretches who had to bring the oil up from the well. Sometimes

Natural oil seep at a mud volcano, Maly Kharami, Azerbaijan.

Natural gas, Yanar Dağ, Azerbaijan. On Sundays, the local people come here to barbecue their *shaslik* in the flames.

The label on this model reads: 'Oil extraction by hand'. Maiden Tower, Baku, Azerbaijan.

the oil would spurt out with tremendous force, tens of metres into the air. These gushers accounted for an important share of oil production until deep into the nineteenth century.

I wanted to discover from Fuad's aerial photos where exactly the oil wells were. Once I had the photos it was relatively easy: in many places the landscape is pitted, as if it has been carpet-bombed. But the pattern is too regular for even the most accurate of precision bombers. There are at least 500 of these wells, and they've all been there since the sixteenth century or longer.

The wells are conspicuously concentrated in two places: Kirmaky Hill to the east of Baku and Yasamal Hill to the west. This is yet another indication of the sensitive noses that people already had for geology many centuries ago: both hills are anticlines, upwardly arched fold structures, like those we found at the Acheron. Oil commonly collects in structures like these. Oil originates in inland seas where there is little water circulation, like the Black Sea today. Plankton living in the seawater sinks to the bottom when it dies. If there is sufficient oxygen in the water, it will decompose, but if there is little circulation, the oxygen is depleted and dead organic material accumulates on the seabed. It is not yet oil, but if sufficient younger sediments are deposited on top of each other, it comes under increasing pressure. The temperature rises, initiating chemical reactions that convert the dead organic matter into oil. Bacteria appear to play a much greater role in these processes than was previously thought. The ideal temperatures for the process are between 50° and 180°C, which are usually found at depths of between 2 and 4 kilometres: this is known as the oil window. If the temperature is too low, the chemical process will not be complete; if it is too high, the oil degrades again, turning first to asphalt and ultimately to graphite.

In the oilfields of the southern Caspian Sea, like those in Azerbaijan, the oil window is much deeper, at 6 to 9 kilometres. This is because here it becomes warmer much less quickly underground than the average for the Earth: an increase of 15°C per kilometre, rather than 30°C elsewhere. Fortunately we do not usually have to extract the oil from that source rock, since oil is lighter than the surrounding rock and rises towards the surface itself. If it encounters no obstacles on the way, we see it emerge as an oil seep. But sometimes there are impervious layers of rock in the way and the oil accumulates in a layer beneath, which is permeable and has high porosity: the reservoir rock. This is often porous sandstone or reef limestone. If the reservoir rocks are folded, the oil naturally rises through the permeable layers towards

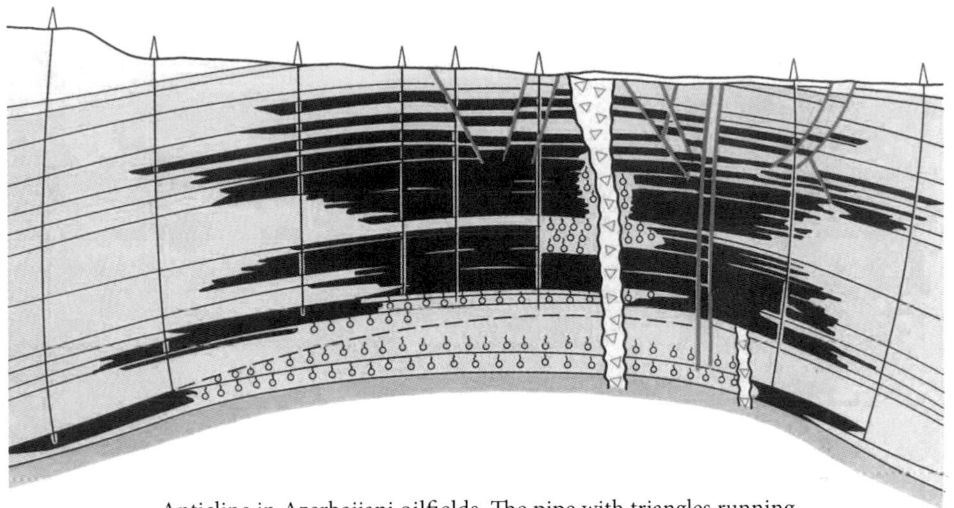

Anticline in Azerbaijani oilfields. The pipe with triangles running through it indicates the supply channel of a mud volcano.

the upwardly arched anticlines, rather than the downwardly arched synclines. The old geologists of the Yasamal and Kirmaky hills were aware of this: both are prominent anticlines.

What is more, Kirmaky Hill is topped with a mud volcano. This is not a coincidence, since the oilfields lie on a package of soft sand and clay containing a lot of water, oil and gas. The reservoir rocks are sand deposited 5 million years ago by the Volga, when it still flowed into the Caspian Sea at Baku. That layer is about 1.5 kilometres thick there, but further to the south it becomes thicker, up to as much as 8 kilometres. The whole soft mass is under high pressure, as here the Arabian and Eurasian plates are colliding with each other. This squeezes the intermediate sediments, forcing them to fold and creating anticlines where the oil collects. Sometimes the pressure becomes so high that the anticline breaks open and the whole cocktail of sand, clay, water, oil and gas is expelled. Try to imagine what would happen if you were to put a lasagne 8 kilometres thick in a vice and tighten the screw. The béchamel sauce would squirt out with a great force.

Sometimes the mud volcanoes are only small pools of mud with bubbles rising to the surface and popping every few minutes, but sometimes they can also cause tremendous eruptions, shooting gas flames tens of metres into the air. The mud volcanoes of the Lökbatan ('drowned camel'), just to the south of Baku, experienced a large eruption in 2001 which created gas flames up to 70 metres high and covered the surrounding slopes with mud flows up to 1.5 kilometres long.

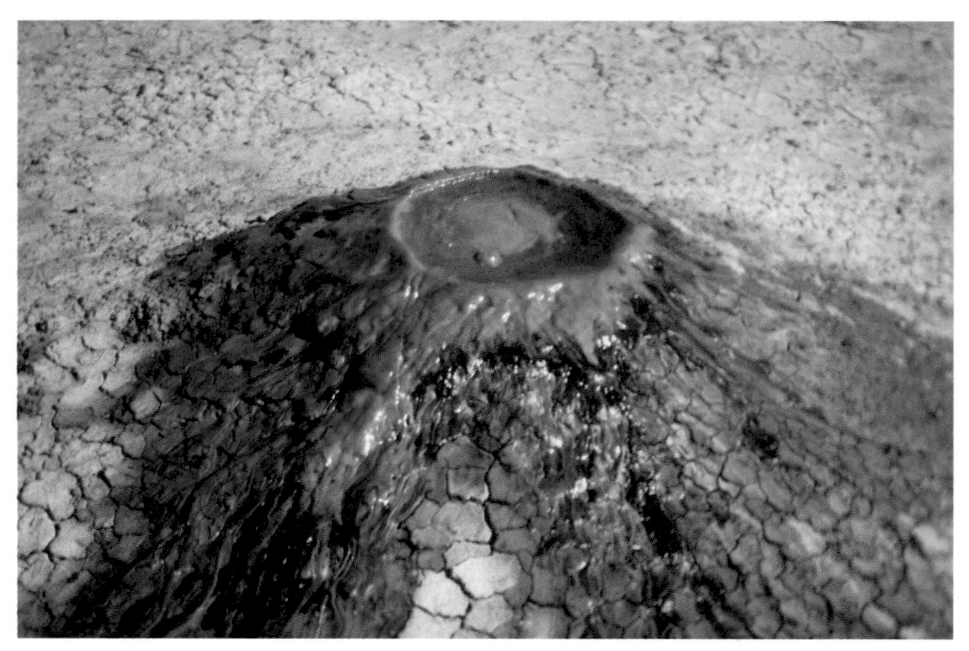

Mud volcano with oil, Akhtarma Paskhali, Azerbaijan.

Mud volcano with active mud flow, Bakhar, Azerbaijan.

Eruption of the Lökbatan mud volcano, Azerbaijan.

Temple to Zoroaster, Surakhani, Azerbaijan, with gas flames burning in the chimneys on the corners.

Mud volcanoes can become almost as large as real volcanoes: up to 400 metres high.

No wonder that, according to legend, the Zoroastrians worshipped fire. Who Zoroaster or Zarathustra was remains a historical mystery. He may have lived in the tenth or the eighteenth century BCE, and it has been claimed in countless locations across the whole ancient Persian Empire that he was born or lived there, including Baku. In the temple of Surakhani at Baku natural gas was piped to the altar for fire worshippers. Fire was not a harbinger of hell, but rather a symbol of purity, justice and truth. '*Il y a peu de religion aussi innocent que ça*', wrote Alexandre Dumas père when he visited the temple in 1858. It reminded him of the Solfatara at Naples, except that this one had the advantage of still being active. Fire appears only to have become part of the Zoroastrian religion at a later date. They did have a hell, the realm of the evil spirit, Angra Mainyu or Ahriman, the 'Lord of the Lies', as Alice Turner translates it so poignantly in her *History of Hell*, with a wink at William Golding. In the final battle between good and evil, hell will eventually be destroyed, purified through burning with glowing metal. In this way ore and oil come together.

I walk up Kirmaky Hill, together with my colleagues Elmira Aliyeva and Dadash Huseynov of the geological institute of Azerbaijan. Fuad's aerial photos now prove their worth: the mud volcano is quiet and there is no longer any oil to be seen in the oil pits. The majority have collapsed, and in a few the remains of the wooden cladding can still be seen. It should actually be a historical monument, as no one extracts oil here anymore. On the other side of the valley it is clear to see why the pits were abandoned. There are the Balakhani drilling rigs. In the early nineteenth century the demand for oil was so great that the old production methods could no longer keep up with it. It was here that oil was extracted by drilling for the first time in 1848 by the engineer V. N. Semyonov, using a rig designed by Nikolay Voskoboynikov. This was eleven years before the United States, which unjustly claims to be the world's first country to drill for oil.

Strictly speaking, this was not the first time that oil had been drilled. The Chinese had been curious about what lay beneath their feet a little earlier. In the eleventh century they drilled for salt in Sichuan using an iron drill bit and a bamboo drill string with a thinner bamboo tube inside it to bring up the salt. Sometimes oil or gas would come with it. In 1512 they reached a depth of 300 metres and, in 1835, as deep as 1,000 metres.

Pit for the extraction of oil by hand, Kirmaky, Azerbaijan;
my shadow shows the scale.

The hill at Balakhani is not a clear anticline, as its fold structure is below ground. But it is part of a longer series of anticlines which are visible on the surface. This, too, should be a historical monument. There is a forest of old, rusty drilling rigs and pumpjacks from the 1940s and '50s: industrial archaeology from the Soviet era. They are still working, though you might wonder how. They seem to be held together with needles and thread. The site is an indescribable mess. Everywhere there are pipes, criss-crossing each other, some connected, others leading nowhere, and puddles of oil, the shallow pools around the rigs pitch black from the oil floating on the surface. Nothing grows here, not only because it is a desert but also because the ground is saturated with oil. On a nearby slope there are a number of bare sagebrush bushes, adorned with plastic bags that have blown against them from an adjacent rubbish tip. Plastic as crop. It is a cheerless site. Elmira, Dadash and I are happy with the few places where the hill has been excavated, because from them we can study the folded reservoir rocks of the old Volga delta closely. But we have to walk across piles of rubbish to get there.

What a contrast to what is happening offshore. Azerbaijan is currently experiencing an oil boom. It is not the first: there was one around 1900, when the country was the largest oil producer in the world. Around the medieval fortified city of Baku there are magnificent mansions from the *belle époque* and a lush park laid out using fertile earth brought back in the holds of empty oil freighters. The Bibi Eybat field on the coast just to the south of Baku was the epicentre of the old oil boom: hundreds of rusty drilling rigs stand there in the oily water, half submerged as a result of the 3-metre rise in the level of the Caspian Sea from the 1970s to the '90s. Alfred Nobel, the Siemens brothers and Henri Deterding made their fortunes here.

Now there is a second great leap forward under way, especially offshore. Baku was the oil capital of the Soviet Union, but the technology used to extract the oil dated back to the 1950s. When Azerbaijan became independent in 1991, it opened up to many Western entrepreneurs, and the country is now known as the Kuwait of the Caspian. Oil had already been sought and found offshore during the Soviet era. It was not actually difficult to deduce that there could be oil off the coast as, during an expedition on the Caspian Sea in 1781–2, Count M. I. Voynovich had, to his surprise, seen the water burning quite a way out to sea: there was a layer of oil on the surface, most probably from an underwater mud volcano. As early as 1803 Haji Kazumbek dug an offshore oil well in the shallow water of the Bibi Eybat field, surrounded with wooden planks to keep the seawater out. It didn't hold for long.

How do you look for oil under water? How do modern oil geologists see the great reserves of oil on the continental shelf, in the North Sea, the Gulf of Mexico and the Caspian Sea? You sometimes need to look though 1 or 2 kilometres of water, and then you're only on the seabed. How can you see that there is oil and gas between 2 and 4 kilometres below that? Fuad cannot show that on his aerial photos; neither can Google Earth. Our eyes just aren't made for looking below the Earth's surface – but we are not completely helpless.

My own objectives are less ambitious. I don't need to look as deep as the oil geologists; what I want to know is how the delta of Azerbaijan's longest river, the Kura, continues under water. That might help us to explain the rapid changes in sea level in the Caspian Sea in the past. But I don't need to look any deeper than 50 metres below the seabed. After my visit to Fuad Mahmudoglu Haji-Zadeh I go to another institute, the KMGRU, in the port of Baku. There are grimy, broken sheets of marble in the corridor, and there's bubbly brown linoleum on the floor in the director's office. But the ship *Geophysics* is moored in the harbour and looks well maintained, which is a good sign. Chart after chart is laid on the table, and courses are set. When can you start? I ask. As far as we're concerned, tomorrow, says the director. And you have to pay in advance, since I have to buy the fuel.

Geophysics, that's the keyword. Geophysics gives us eyes under water. Whether you want to look down 50 metres, 5 kilometres or right through the whole Earth, the principle of seismic mapping is the same. From a boat you create a shock wave in the water, for example by using an air gun to blow a large air bubble in the water. The bubble then pops suddenly, and the shock wave it causes spreads out through the water until it reaches the seabed. Part of the shock rebounds back to the surface of the water, while another part penetrates into deeper layers below the seabed. At the boundary plane between two layers, the wave can rebound to the surface of the water, but will of course arrive later than the parts of the wave that rebound directly from the seabed. Behind the boat a line of hydrophones is suspended in the water. They pick up the signals and the arrival times of the different signals from a continuous series of shock waves, producing a two-dimensional image of the layer structure of the seabed. The result often looks like bad knitting. Since offshore oil drilling can easily cost tens of millions of dollars, oil geologists prefer to make a seismic map before they start.

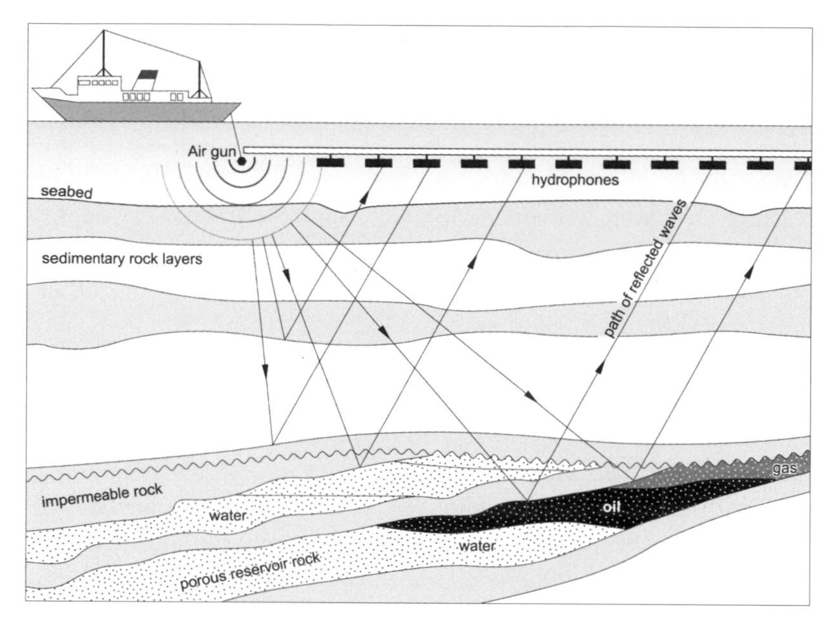

The principle of seismic exploration.

The frequency of the shock wave determines how deep you can look. With high-frequency exploration methods, you can see very fine layers, but you cannot penetrate further into the seabed than a few metres. That is not enough for oil geologists. As we have seen, oil-bearing layers usually lie between 2 and 4 kilometres deep, and for that you need to create low-frequency waves. The picture (on page 220) may then be much less detailed, but formations like subterranean anticlines will show up clearly. What oil geologists cannot see, with the occasional exception, is the oil itself. They can only see whether the formations are favourable, and then they have to use their knowledge of source and reservoir rocks to decide whether to drill.

Seismic exploration was conducted during the Soviet era, but the technology was not very advanced. Better recording techniques, better interpretation software and three-dimensional seismology have enormously improved the results. A whole network of seismic lines is laid over an area where good results are expected. The lines are drawn like boustrophedon text, from the classical Greek meaning 'back and forth, as the ox ploughs'.

The seismic data obtained can then be used to construct a three-dimensional image of the structure of the underground. One can, as it were, take a virtual walk through the oilfield in what is known as a 'cave', a kind of theatre where one can use 3D glasses to see in three dimensions.

It is also possible to plot all kinds of data collected during drilling, such as the electric conductivity of the rock, the radioactivity (which tells you how much clay there is in the rock), the chemical composition, and many other well-logging methods. In this virtual world you can walk along old river courses of the Volga, through old coral reefs and subterranean mud flows, until the 3D makes you nauseous. These new techniques have dramatically increased the success rate of oil drilling, from 20–30 per cent to 50–60 per cent. In Azerbaijan, too, oil has brought incredible wealth to the country.

Baku is no longer grimy. The splendid medieval city centre has been renovated and the *belle époque* houses restored. The promenade along the seafront is a true pleasure, and the old Soviet hotels, with their creaking parquet floors and threadbare carpets, rusty taps and narrow, sagging beds, have been demolished to make way for the Hilton, the Marriott and other luxury but nondescript chain hotels. Many 'Khrushchoby', ramshackle white brick housing blocks from the Khrushchev era (*trushchoby* means 'piles of rubbish', hence the name), have been replaced by glittering glass palaces. The one-man businesses selling pomegranates and spices are now shops with international names like Lanvin, Estée Lauder and Benetton. The Azeri in the street cannot afford to shop there, only those who should actually be splashing around in Dante's river of tar. You don't get clean by bathing in riches.

I did not go back to see Fuad at his institute, but I could see from its website that it has undergone an impressive process of modernization. The geological institute has set up a hypermodern laboratory for Earth observation and geographical data processing, even though most of the staff are still pensioners earning a little on the side. When I returned to the KMGRU in 2007 for a second seismic exploration, the director was no longer able to start the following day as he had so much work to do for the oil companies. I had to wait a year and a half. And the floors no longer creaked.

The most impressive technological improvements have taken place in drilling itself. One of the problems is that you cannot easily anchor a drilling rig to such a soft seabed: there is a serious risk of it shifting around under the water. I asked ExxonMobil geologist David Puls, who has worked in Azerbaijan for many years, about the stability of the seabed. 'Seabed stability?' he replied, in a sepulchral voice, 'There isn't any.' BP geologists compared the seabed with yoghurt. The Caspian Sea has now become a technological test area for drilling

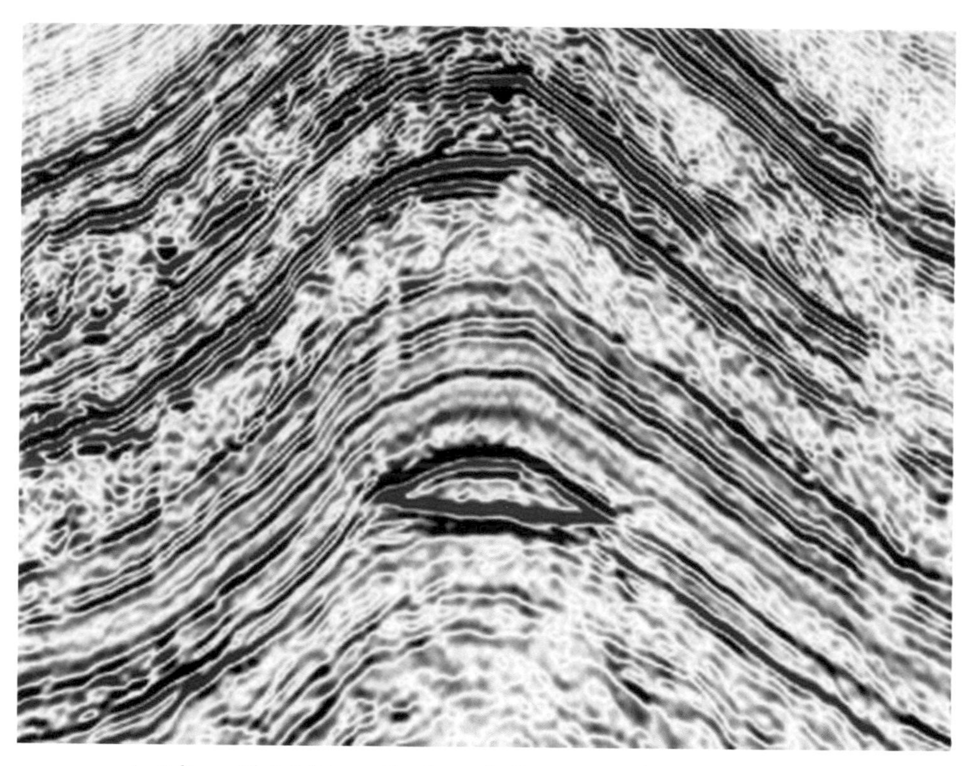

Anticline with bright spot (horizontal oil–gas contact) in a seismic profile.

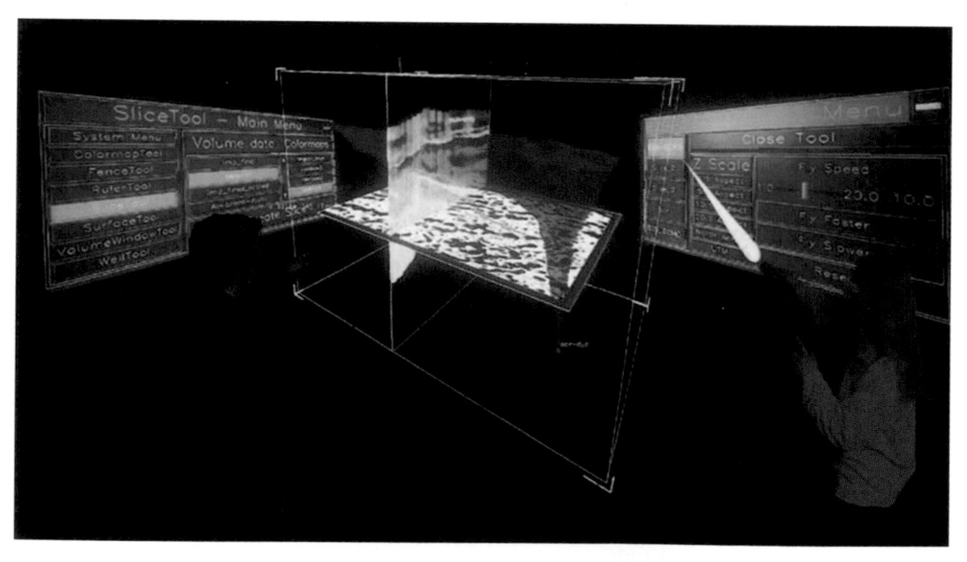

3D visualization of seismic data.

The Shah-Deniz-Alpha TPG-500 offshore production platform.

under such difficult circumstances. Part of the solution is floating oil platforms, which have to be built on the spot since, as the Caspian is an inland sea, you can't drag a drilling platform there from somewhere else.

What lies *under* the seabed is even more capricious. Because the mud is under high pressure, you have to drill very carefully. During drilling in the Shah Deniz gas field, one of the largest in the world, the pressure suddenly dropped: it was as though the drill bit had fallen into a hole. It was probably the reservoir of an undersea mud volcano that had very recently emptied. A sudden loss of pressure during drilling also occurred in a completely different part of the world, in the offshore Nang Nuan field in Thailand. But the cause was different: they had stumbled across an oil-filled karst cave in Jurassic limestone, 3 kilometres under the seabed. I wonder if anyone, at the Shah Deniz or the Nang Nuan fields, thought for one moment that they had drilled through the roof of hell?

After all, it is very dangerous to drill a hole in soft lasagne that is under high pressure: you can easily cause a gusher. There is technology to prevent that happening onshore but offshore, and especially in deeper water, the risks are still enormous, as the explosion of the Deepwater Horizon drilling platform in the Gulf of Mexico in 2010 shows. For the eleven workers who died in the incident, it was indeed hell. 'Birds were falling out of the sky', wrote *Fortune* magazine in a reconstruction of the disaster. Wikileaks has revealed that a similar accident occurred in September 2008 in one of the largest oilfields in Azerbaijan, the Azeri-Chirag-Günesli.

An additional problem in the south Caspian Sea is that the oil is at such a great depth. In the Zafar Mashal field oil company ConocoPhillips drilled more than 6 kilometres below the seabed in water 600 metres deep, and still had to cease the drilling operations before reaching the oil-bearing reservoir rock because of technical problems. Six kilometres! This is deeper than anyone has ever been physically, deeper than the deepest mine in South Africa. And it is not even the greatest depth at which oil has been drilled: in 2009 the ill-fated Deepwater Horizon drilled down more than 10 kilometres in the Gulf of Mexico.

Once oil or gas has been found, it is inconvenient to have to construct a new platform every time the drilling is moved to a new place. A lot of companies are therefore increasingly drilling at an angle, or even horizontally, from the same location, resulting in a whole forest of boreholes on the seabed. The technology is now so advanced that, if water suddenly flows into a drill bit instead of oil, for example because the oil in the layer has been depleted, a sensor in the bit automatically sends a signal to the drilling platform. This is then transferred to the headquarters on shore, where engineers can decide whether to close off the bit by remote control. The drill-bit also has a valve that can close automatically to stop water from flowing in.

Via sensors on the seabed, it is possible to obtain repeated 3D seismic images, so that geologists can follow how the field changes as a consequence of the oil and gas production: this is known as 4D or time-lapse seismology. One of the more striking findings of this kind of monitoring is that oil extraction has caused the seabed in the Ekofisk field in Norway to subside as much as 10 metres. Submarine subsidence: hidden from sight, there are all kinds of changes happening, kilometres under the beds of the world's oceans.

In the oil and gas fields of the Caspian Sea what is often extracted is gas condensate: a mixture of natural gas, light hydrocarbons and hydrogen sulfide, H_2S. Once again, the sulfur is produced by bacteria that have reduced it from the sulfate in the seawater. But no one wants it. At the Tengiz field in Kazakhstan in the northeast of the Caspian Sea, so much excess sulfur is produced that areas as large as housing estates are being filled up with stacks of sulfur blocks.

Deep drilling also supplies unique scientific data. There is an international scientific programme that is trying to drill as deep as possible into the Earth's crust. One of these deep continental drilling sites is in Azerbaijan, on land near Saatly in the valley of the Kura river. The drilling penetrated to 8,324 metres but produced little information that is new. It reached Jurassic volcanic rock that can also be found on the surface in the northeast of the Lesser Caucasus.

The deepest hole ever drilled was in the Kola peninsula, in the extreme northwest of Russia, close to the border with Norway. It is an interesting location because it is the place where the oldest rocks in Europe occur, some 2.5 to 3 billion years old. It is logical to think that if you drill into them, you will come across even older rocks. Geophysical data also suggested that, at greater depths, the Earth's crust would have a different composition than closer to the surface. The goal was to drill down to 15 kilometres. The project started in 1970 and reached its maximum depth of 12,262 metres in 1989. It couldn't go any deeper because the temperature down there was 180°C, 80°C higher than the team had expected. At 15 kilometres deep, the temperature would have increased to 300°C. The drilling equipment was not capable of dealing with such high temperatures. The team did not find the expected transition to older rock but were very surprised that the rock at that depth was full of water, apparently from the depths of the Earth. The oldest rock they found was 2.7 billion years old, but similar rock can be found on the surface, just over the border in Norway.

When the Kola borehole reached its deepest point in 1989, the Christian Trinity Broadcasting Network in the US reported that scientists somewhere in Siberia had drilled through to hell. The report was published in all kinds of obscure newspapers and magazines. They claimed that, at 14.4 kilometres deep, the drill started to rotate wildly, and the temperature had increased to 2,000°C. A microphone lowered into the hole was alleged to have picked up the screams of hundreds, perhaps thousands, of doomed souls. The project leader, Dimitri Azzacov, claimed that the Earth was hollow.

It was a classic urban legend: there was no geologist called Azzacov, the deep drilling took place in Kola, not Siberia, and the depth and the

Drilling at an angle from a production platform, Ekofisk field, Norway.

temperatures were fictitious, not to mention the story about the microphone. American journalist Rich Buhler tried in vain to discover where the report came from. But for some Christians, it was proof that hell – and therefore God – existed. Microphone or no microphone, they heard what they wanted to hear.

Sulfur storage from gas condensation, Tengiz oilfield, Kazakhstan.

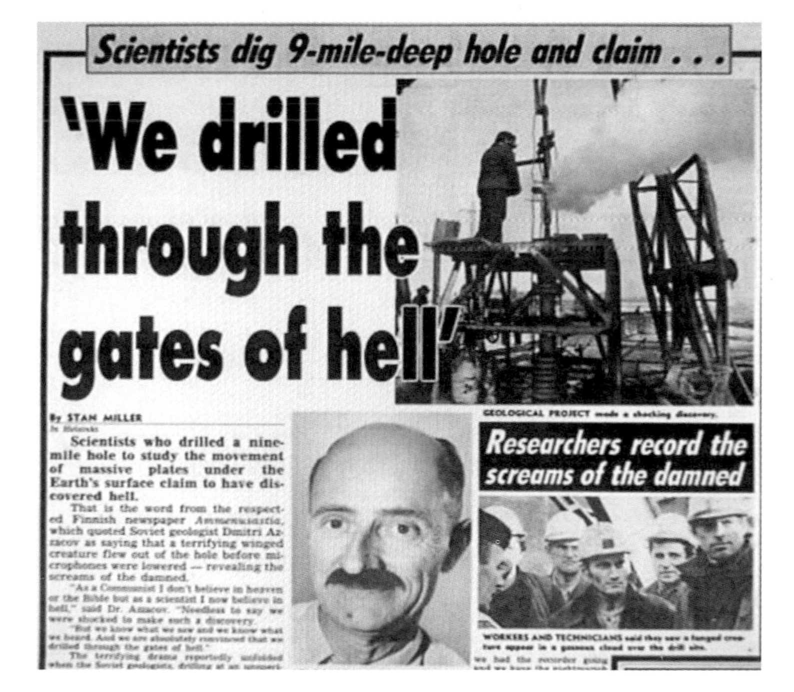

Scientists dig 9-mile-deep hole and claim . . .

'We drilled through the gates of hell'

By STAN MILLER

Scientists who drilled a nine-mile hole to study the movement of massive plates under the Earth's surface claim to have discovered hell.

That is the word from the respected Finnish newspaper *Ammennusastia*, which quoted Soviet geologist Dmitri Azzacov as saying that a terrifying winged creature flew out of the hole before microphones were lowered — revealing the screams of the damned.

"As a Communist I don't believe in heaven or the Bible but as a scientist I now believe in hell," said Dr. Azzacov. "Needless to say we were shocked to make such a discovery.

"But we know what we saw and we know what we heard. And we are absolutely convinced that we drilled through the gates of hell."

The terrifying drama reportedly unfolded when the Soviet geologists, drilling at an unspeci-

GEOLOGICAL PROJECT made a shocking discovery.

Researchers record the screams of the damned

WORKERS AND TECHNICIANS said they saw a fanged creature appear in a gaseous cloud over the drill site.

we had the recorder going

'Well to Hell', article in the American supermarket tabloid *Weekly World News*, 24 April 1990.

THIRTEEN

Collapses

Not only was that place, where we had come
to descend, craggy, but there was something there
that made the scene appalling to the eye.

Like the ruins this side of Trent left by the landslide
(an earthquake or erosion must have caused it)
that hit the Adige on its left bank,

when, from the mountain's top where the slide began
to the plain below, the shattered rocks slipped down,
shaping a path for a difficult descent.
Dante Alighieri, *Inferno*, XII:1-9

Have we have started to dig large holes everywhere horizontally from pure frustration that we have been unable to go down deeper than 12 kilometres? We have dug not just narrow holes to get at oil, but also enormously long horizontal tunnels. At 57 kilometres, the new Gotthard Tunnel under the Alps is the longest tunnel in the world. It is 2 kilometres under the highest summit. If it had been drilled down 57 kilometres vertically, we would have seen a lot more! But it is still a technological masterpiece. The Roman tunnels in the Neapolitan tuff at Lago Averno, and even modern mine tunnels, are child's play in comparison.

The story of modern tunnels started a long time ago. The French-British engineer Marc Brunel was the first to dig a tunnel under the River Thames in London in the early nineteenth century, using an ingenious system of movable shields which dug away the weak London clay a little at a time. It was not a simple task. Filthy sewage sludge seeped through the roof; methane gas exploded in the tunnel; the river repeatedly broke through, flooding the tunnel and drowning people inside; and, once it was finished in 1841, it became largely a place of refuge

Room and pillar system in an underground basalt mine, Mendig, Germany.

for the homeless, thieves and prostitutes. It was only possible to walk through it; it was even impassable with a horse and wagon. But it was a start. The tunnel is now part of the London Underground.

When the Paris Metro was built at the end of the nineteenth century, the tunnelers discovered anything but virgin ground beneath the city. Just like Jerusalem and Naples, Paris has its negative. Since the time of the Romans it has been built of rock dug out from below the city itself. An estimated 10 per cent of the ground under the city is hollow. The subterranean quarries for Lutetian and Cretaceous limestone for building, and for gypsum for plaster of Paris, cover a total area of 3,000 hectares. The most commonly used mining system was room and pillar, which entailed leaving columns of limestone or gypsum to support the ceiling. Sometimes, however, there was too much room and not enough pillar and the roof would collapse. The wonderful fossilized mammal bones that Georges Cuvier removed from the gypsum mines paved the way for his scientific success. Here, too, there are subterranean rivers in the limestone, water supply channels, sewers, small dripstone formations, mushroom nurseries and catacombs containing the remains of an estimated 6 million Parisians. The Metro had to find its way through all of this. Brunel's shield system was also tried in Paris, but the limestone, marl and sandstone proved too much for it and they had to fall back on classic mining techniques with wooden struts in the tunnels.

During the building of the metro in Moscow in the 1930s, the workers continually encountered quicksand. Covered with moraines from the ice ages, there were treacherous, deep, sand-filled river valleys that immediately emptied out when they were broken open. The tunnels were not drilled: everything was dug out using spades, pickaxes, wheelbarrows and buckets. The quality of the concrete was a problem, the workers were unskilled, the pressure of work was high, coordination was inadequate and they continually had to hunt down saboteurs and *kulaks*. Many galleries collapsed, writes Josette Bouvard in her book *Le Métro de Moscou*. On Lubyanka Square, near the notorious prison, a tunnel collapsed, leading the nearby station Dzerzhinskaya to be built without a central hall. Later, during the building of the Borovitskaya metro station in the centre of the city, the famous Lenin Library was damaged. At the opening of the first line on the seventeenth anniversary of the October Revolution in 1934, chief overseer of the project and Stalin's right-hand man Lazar Kaganovich said: 'We fought with nature, with the poor soils beneath Moscow. Moscow's geology turned out to be pre-revolutionary, old regime. It worked against us.'

This all compellingly calls to mind the subterranean collapse witnessed by Dante and Virgil when they descend from the sixth to the seventh circle. The landslide in Trente with which Dante compares it is still a prominent feature of the landscape near Lavini di Marco. And in the eighth circle, in the Malebolge, they come across another collapse.

> He answered: 'Closer than you might expect,
> a ridge jutting out from the base of the great circle
> extends, and bridges every hideous ditch
>
> except this one, whose arch is totally smashed
> and crosses nowhere; but you can climb up
> its massive ruins that slope against this bank.' (XXIII:133–8)

It is November 2009. Amsterdam is digging tunnels for a new North-South metro line. Gigantic drills with a digging disc of 8 metres in diameter, the size of a whole tunnel tube, will chew their way through the soft ground at a rate of 20 metres a day. That is a risky undertaking because most of the houses in the historic city centre are built on piles, driven through the soft clay and peat in bygone centuries into

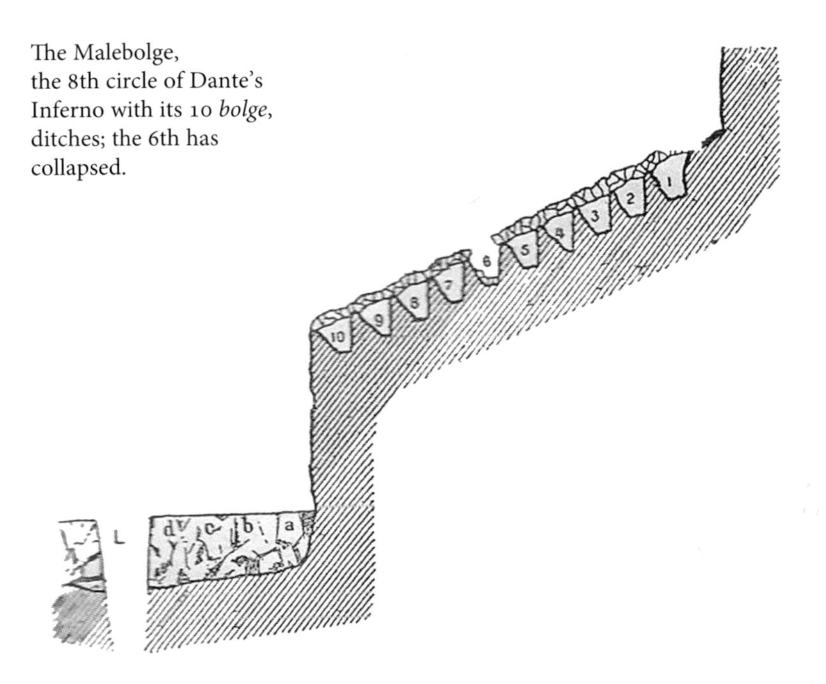

The Malebolge, the 8th circle of Dante's Inferno with its 10 *bolge*, ditches; the 6th has collapsed.

a more stable layer of sand. The piles must not only not be disturbed but also must not be allowed to rise above the groundwater, otherwise they will rot and lose their strength. On one of the canals, the Vijzelgracht, a few houses have already subsided and developed cracks as a result of leak in a retaining wall.

Johan Bosch, chief engineer of the North–South line project, invites me to take a look at the construction site. That is an opportunity not to be missed, and I jump at the chance to see the city I studied in from below the ground. Johan leads me past the fences and down the steps to 15 metres under the Rokin, and we are in another world, away from tourists and trams. There, in what will become an underground station and car park, an immense space is being kept open with imposing steel retaining walls, supported by horizontal round steel beams. The drilling hasn't started yet; that will come later.

The reason I am so keen to go down there is not to see the retaining walls, but the one place where there are no walls, at the extreme north end of the construction pit, where a piece of Amsterdam's underground geology is visible. At the bottom is the layer of sand on which the piles of the city's houses rest. On top of that, with a very clear dividing line, there is a thin deposit of peat and finely stratified clay. This is the renowned basal peat! I walk to the outcrop, feel the sand, the peat, the clay, wheedle out delicate shells, see burrows dug by

St Barbara is asked for her blessing during the initiation of the tunnel drill in Amsterdam.

subterranean creatures, and take photographs to my heart's delight. It is a moving experience.

Master Pieter Pieterszoon Ente was the first to see this peat – almost 400 years ago, in 1605, during the digging of a water well at the Oudemanhuispoort. It was the world's first scientifically recorded drilling operation. As he described it himself,

> On the first day, they drilled up 51 feet of earth, mire and peat, soft clay, sand and somewhat harder clay. On the second day, they passed through a layer of 22 feet of sand, on which most houses in Amsterdam are built; and then through blue clay, white sand, loam and humus. On the third day, they dug through 14 feet of pure sand, and in the three days that followed they penetrated a further 55 feet deeper. First they encountered sand with clay, then sand with seashells and snail shells, followed by hard clay, sometimes pure but more often mixed with small shells. In the subsequent seven days, they dug through 62 feet of mostly hard clay and, lastly, through 13 feet of sand, mostly mixed with small stones. In the final eight days, they dug through 28 feet of pure sand.

The basal peat and the first marine sediment on top of the Pleistocene sand layer, Rokin, Amsterdam.

Since an Amsterdam foot is equal to 0.283 metres, according to Ente the transition from 'mire and peat' to sand is at a depth of 14 metres, quite close to what we saw on the Rokin. This description became so famous that it is quoted literally in a Russian book from 1763, *O sloyach zemnykh* (About the Layers of the Earth) by Mikhail Vasilyevich Lomonosov, founder of the Moscow State University – without the source being credited.

But what do that sand, peat and clay actually mean? As a geologist, you not only want to see what these layers look like, but also know what it is you are looking at. The sand was on the surface during the last ice age, around 20,000 years ago. During this period the Netherlands was a deep-frozen polar desert, and the North Sea was dry because the sea level had fallen dramatically, by some 120 metres. There was no ice cap – that was much further to the north, in Denmark and northern Germany. Hard ice-cold winds around the ice cap whipped up the sand from the dry seabed and deposited it as cover sand and snow dunes. That is the lowest layer of sand, and is not particularly special, since the same sand can be seen on the surface in many parts of the eastern Netherlands. But here in Amsterdam it is buried under peat and clay.

The peat does not date from the ice age but from the warmer period that followed it. It tells us that the sea level has risen in the past 8,000 years as a result of the melting of the ice caps. If the sea level rises, the groundwater below the land rises along with it. That creates wet areas on the surface, where peat will eventually develop. The higher the sea level rises, the further inland the peat will be found. The clay layer with shells that lies on top of the peat shows how the sea flooded the land once the level became high enough. If you can determine the age of the peat, you can also determine how much the sea level has risen in the past 8,000 years.

The Netherlands was the first country to research this phenomenon, making the basal peat famous around the world. The research was described in a thesis by Saskia Jelgersma of the Dutch National Geological Survey in 1961. Previous attempts to explain it had used grains of pollen, which do not enable you to determine absolute age, so the results therefore remained rough estimates. Jelgersma was the first researcher in the world systematically to date the levels in the basal peat in various bores in a cross-section of the western Netherlands, using the carbon-14 dating method developed in 1949. The results showed that the sea level rose rapidly at first and then more slowly, and that the simultaneous subsidence of the ground was slower in the southern province of Zeeland than further to the

Fossilized shells in the basal peat, Rokin, Amsterdam.

north. Jelgersma's research was repeated at the same locations in 1982 by Orson van de Plassche of the Free University of Amsterdam and Henk Berendsen of Utrecht University in 2007. The results showed that, despite all improvements in dating techniques in the past 40 years, Jelgersma's curve is still very reliable. Such historical information is very important in allowing us to see predicted rises in sea levels in the future in the light of past developments.

What a great shame that these engineers were going to fill this unique piece of the underground of the Netherlands with concrete, allowing no one ever to see it again. It is a unique piece not only of geological history, but also of scientific history. The basal peat and Saskia Jelgersma deserve a monument. Fantastic data were also collected when they built the Velsertunnel in the 1950s. There the basal peat lies 20 metres deep. It is the oldest in the country, more than 8,000 years old, but you cannot see it now. And all the material they have removed from this pit in Amsterdam is now irrevocably lost to the study of the city's geological history. What has been excavated was an enormous record of the underground. The Earth is its own history book and we have only one copy. Imagine that there was only one copy of *On the Origin of Species* and an army of bookworms chewed holes in it: part of the information would disappear forever. In the Netherlands, over the course of time, some 400,000 boreholes have been made. You can access the results on the Internet but in most cases, you can't study the material yourself.

And where has all the ground excavated for the North–South line gone? It has been crushed, ground and mixed together until it is unrecognizable, and has probably been used to build new islands elsewhere in Amsterdam. It has become a new raw material, from which all traces of its earlier history have been erased. Someone has pressed the reset button. It reminds me of all those poor amateur geologists who eagerly sift through the sand deposited to create Maasvlakte 2, the new artificial port built on reclaimed land near Rotterdam, searching for fossils that have been sucked up with the sand from the bottom of the North Sea. No one will ever know in which layer the fossils originated.

In actual fact, all minerals that are extracted from the ground leave a hole behind them. The iron in the beams supporting the underground car park on the Rokin left a hole in the iron mines of Australia or Brazil. The concrete that covers the basal peat so unceremoniously comes from the limestone quarries of Limburg or Belgium, and the gravel that

Mammoth bone found in sand used to create the Maasvlakte.

is mixed into it from the gravel pits of the Rhine and the Maas. Holes everywhere. The oil used to drive the generators, the asphalt on the road of the Rokin, the plastic sheeting to cover everything up, is all sucked out of oil and gas fields. The Vinkeveense Plassen and the Loosdrechtse Plassen, artificial lakes in the Netherlands that are popular venues for water sports, are actually depleted peat pits. The holes created by our coal and marl mines are still causing subsidence, as are the gas fields in the northern province of Groningen and the salt caverns in Friesland and Twente. To meet its energy needs, the Netherlands has practically hollowed itself out. It is a miracle that we haven't all disappeared into a big hole. Only wood leaves no holes behind, except a minimum reduction in the CO_2 level in the atmosphere, and that is only temporary. If it's not grown, it's mined.

In 2009, during the construction of the metro in Cologne, the city's Historical Archive collapsed. The incident not only caused the death of two victims, but also meant the loss of unique records of more than 1,000 years of history. The people of Cologne had removed the history of the Earth from under the city and the Earth took its revenge by swallowing up their history.

Dinosaur bones in the Jurassic Morrison Formation, Dinosaur quarry, Utah.

FOURTEEN

The Lead Cloak

Dazzling, gilded cloaks outside, but inside
they were lined with lead, so heavy . . .
Dante Alighieri, *Inferno*, XXIII:64–5

The Kola borehole was 12 kilometres deep, the deepest we have been able to penetrate with our drills. Twelve kilometres, while we have known for more than 2,000 years that the distance from the surface of the Earth to its centre is more than 6,000 kilometres. So we haven't got very far; we've only actually been able to see a quarter of 1 per cent of the inside of the gobstopper with our own eyes. In the introduction to this book I promised that we would go a little deeper in each chapter. And here we are nearly at the end, with almost the entire journey ahead of us.

From now on we will have to make do with indirect observations, courtesy of geophysics, and seismology in particular. During the search for oil and gas, as we have seen, a shockwave is generated that partly rebounds, partly penetrates through to deeper layers and then rebounds again. But the best way to look through the underground of the Earth is using natural shockwaves, caused by earthquakes.

Croatian scholar Andrija Mohorovičić was originally a meteorologist, but from 1901 he devoted himself entirely to seismology, 'to investigate the interior of the Earth and to take over where the geologist stops, because the modern seismographs can serve as a binocular for observing even the greatest depths'.

There are two main types of earthquake waves: a primary (P) wave characterized by alternating compressions and rarefactions that move in the direction in which the wave is propagating; and a secondary (S) wave which moves perpendicular to the direction of the wave's propagation. The waves are called primary and secondary because P-waves always arrive at a seismograph on the Earth's surface earlier than S-waves.

In 1910, in a Croatian article with the humble title 'Potres od 8. x.1909' (The Earthquake of 8 October 1909), Mohorovičić described how the P- and S-waves of an earthquake in the Kupa Valley in Croatia arrived at the seismograph in two distinct pairs. From that he deduced that there must be a marked structural discontinuity below the Earth's crust. We know now that this discontinuity, later named the Mohorovičić Discontinuity (Moho for short), forms the underside of the Earth's crust. In the Netherlands it is at a depth of 27 kilometres, under the oceans at 5 to 10 kilometres, and under the Andes and the Himalayas at 70 kilometres. The Earth's crust is thicker under mountains than under flat land, just as a higher iceberg lies deeper in the water, due to the principle of isostasy. Geodesists, especially George Airy and John Henry Pratt, had already deduced the principle in the mid-nineteenth century on the basis of gravitational measurements. They suspected that the continents floated on denser material, though they had no idea what that material was. Mohorovičić was the first to show that there is indeed a discontinuity between the lighter crust material and the heavier mantle.

In the 1920s the Austrian-American seismologist Victor Conrad discovered another discontinuity, this time within the Earth's crust. He thought that it was the dividing line between the slightly lighter upper crust of sial (silica and aluminium-rich rock, like granite) and the somewhat heavier, basaltic lower crust of sima (silica and magnesium-rich rocks, like basalt). It was the Conrad discontinuity that they were looking for during the super-deep drilling to a depth of 12 kilometres on the Kola peninsula, but without any luck – so far at least.

The mantle consists mainly of peridotite, named after its main component, the green mineral olivine or peridot. No one has yet succeeded in extracting these rocks from the deep interior of the Earth. But just as the fold formation in the gorge of the Acheron in Greece has given us a preview of deeper layers, the Earth itself has pushed upwards what we have been unable to find ourselves by drilling. Many basaltic volcanoes, like those in the Eifel region of Germany and the Auvergne in France, have dragged up fist-sized chunks of peridotite and similar rocks from the depths of the outer mantle. There are even whole massifs of peridotite or related rocks, like serpentine, that originate from the Earth's mantle.

It is one of the stories I remember the most clearly from the lectures by Professor De Roever, my later PhD supervisor and leader of our expedition to the copper mine in Falun. He was an excellent

Ronda peridotite in southern Spain, in the foreground.
In the background are Triassic marbles.

teacher, but this story stuck in my mind the longest: it was about a discovery he himself had made. In 1957 he had published an article in the German journal *Geologische Rundschau* entitled 'Sind die alpinotypen Peridotit-massen vielleicht tektonisch verfrachtete Bruchstücke der Peridotitschale' (Are the Alpinotype Peridotite Massifs Perhaps Tectonically Transported Fragments of the Peridotite Mantle?). He had always reproached himself for publishing the article in German, as a result of which it may not have received the attention it deserved. Perhaps he was too hard on himself: it was included in an English compilation of pioneering articles on his discipline, petrology, and is still regularly quoted to this day.

De Roever had seen such peridotite massifs on Celebes in the 1950s and concluded that they had could not have crystallized out of magma, but had been thrust upwards as slices of hard rock during intensive folding of the mountains. But where from? His groundbreaking idea that they had come from a great depth is now generally accepted, while at the time the concept of plate tectonics had not yet been widely embraced. The upshot is that we can study the mantle rocks at many places around the world without ever having to go deep underground. Mantle rocks like peridotite and serpentinite are often used as natural stone cladding for shop fronts, so you don't have to make the effort to go and see them in their natural surroundings. Even green asbestos is originally a mantle rock.

De Roever's enthusiasm aroused in me a desire to go further in petrology. In more than six years with the Geological Mining Service in Suriname I was able to study rocks that may not have come from the mantle, but certainly from very deep in the Earth's crust. All things being considered, large parts of the current surface of the Earth are comprised of rocks that originated several kilometres, or even tens of kilometres, deep in the Earth. Granite, for example, generally crystallized some kilometres deep and is only found on the surface because the layers on top of it have been eroded away over the course of millions of years.

I collected rocks alongside the Courantyne and Coeroeni rivers in the southwest of the country, with a field team of twenty Surinamese labourers who work in the forest, paddling down the rivers in dug-out canoes or walking through the virgin rainforest for days on end, carrying a basket on my back with a hammock, primus stove and a zinc chest for my clothes and food. I collected samples: beautiful gneissic rocks with light veins and dark bands. It was almost impossible to imagine that they had all once been clay, but the presence of certain aluminium-rich minerals proved that that was once the case. Two billion years ago the clay was deep in the Earth's crust, and was even partially molten: the white bands showed where it had melted.

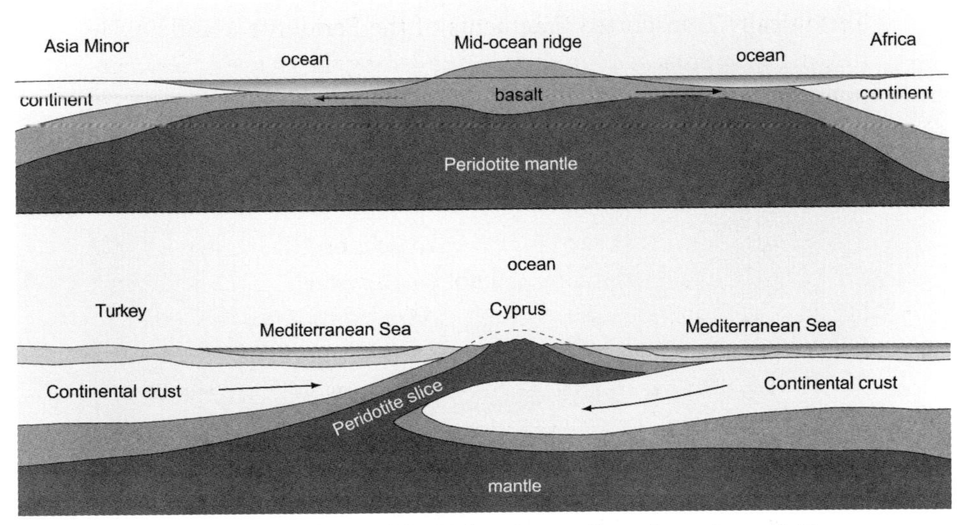

During the contraction of the Tethys Ocean, slices of mantle peridotite were thrust to the surface where Cyprus now lies.

Collecting rock samples in Coeroeni, Suriname. The rocks were hot as hell.

Gneiss, which was once clay. Worsteling Jacobs – Sara's Lust plantation, Suriname.

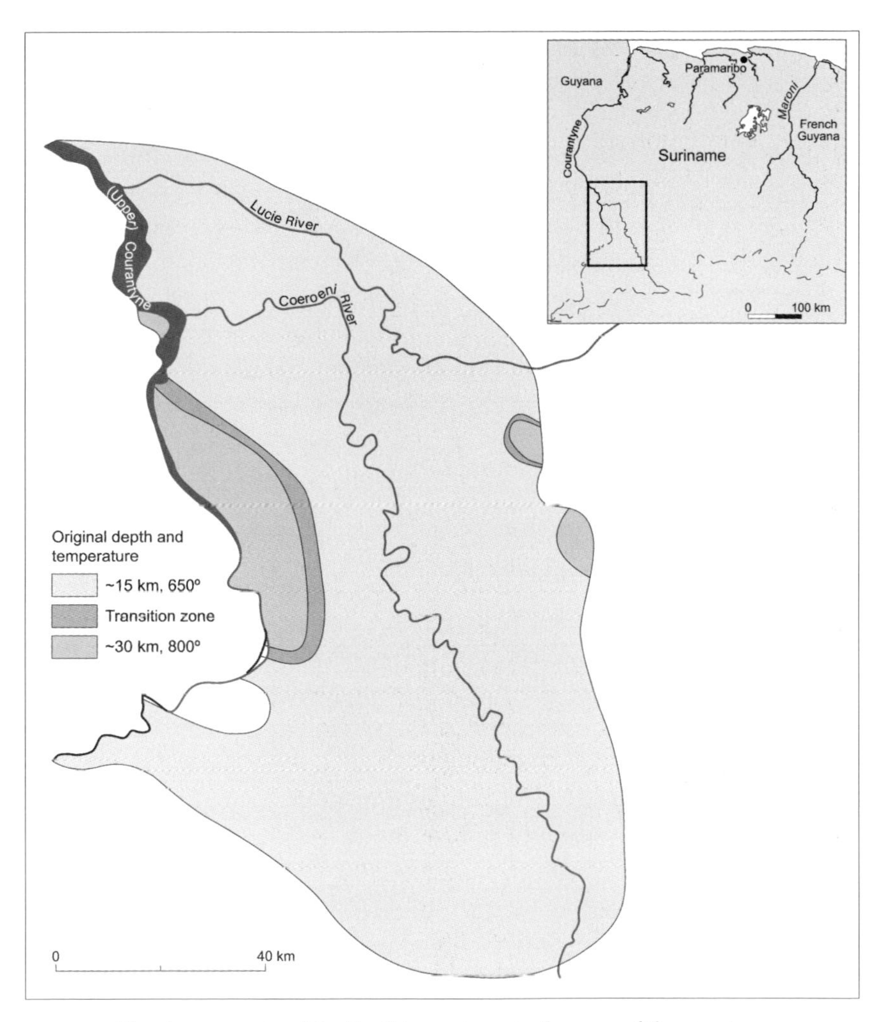

The deeper parts of the Earth's crust are in the core of the structure.

I carefully mapped where those minerals were to be found, studied the rocks under the microscope and discovered, as we paddled southward down the Courantyne, that the white mica muscovite suddenly disappeared from the rocks.

Other minerals took its place, in particular large crystals of red garnet and blue-grey cordierite and small ones of green spinel. There were also all kinds of changes in the darker rocks, which were probably old basalt lavas. Paddling along the Courantyne, I had penetrated a little deeper into the Earth's crust: the rocks with muscovite had been formed at a depth of about 20 kilometres at temperatures around 650°C, those with garnet at 30 kilometres and 800°C. Laboratory tests showed that muscovite is no longer stable at such higher pressures and temperatures, while other minerals are. From that perspective, I have been 30 kilometres deep in the earth's crust, much further than the 700 metres in Kiirunavaara. But it wasn't much like hell; more like the Elysian Fields.

Perhaps I could have gone even deeper, but that would have been too risky: the most interesting rocks lay in the triangle between the Upper Courantyne and the Coeroeni, which is claimed by both Suriname and Guyana. Since an attack on the Surinamese border post at Tigri in 1969, the Guyanese have controlled the area. I was able to take a few samples there, but I think I was probably very close to hell.

I've found something really interesting, I wrote to De Roever. He thought it was interesting too and wrote back that I could write my PhD on it. At a depth of 30 kilometres, he mused, we might be below the Conrad Discontinuity; he told me to look to see whether there were enclosures in the minerals containing CO_2. An article by Jacques Touret, later a professor at the Free University of Amsterdam, had suggested that this might be the case as, when the mantle releases gas, a lot of CO_2 rises up through the Earth's crust. Unfortunately, nothing came of that new research, as my contract in Suriname expired.

We can look even deeper into the mantle at hotspots, plumes of rising mantle material originating from far below the moving lithospheric plates. The most well-known example is Hawaii. The hotspot that currently lies beneath Hawaii remains pretty much stationary, but the Pacific plate above it is sliding over the top of it. As a result, in the past, the hotspot has burned a whole series of volcanoes through the plate.

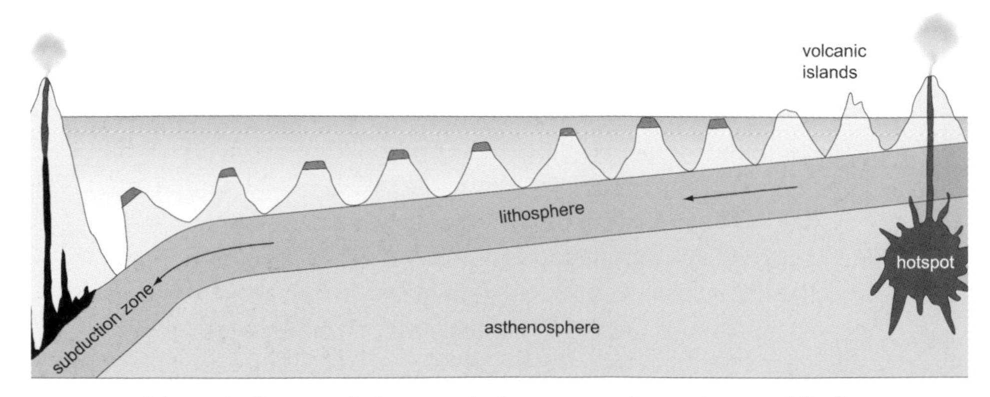

Schematic diagram of a hotspot: the hotspot remains stationary while the lithospheric plate above it moves over it, creating a whole chain of volcanoes.

Fire and ice on Iceland: volcanic ash from the eruption of Eyjafjallajökull in April 2010 on the Gig glacier on the northern side of the volcano – the ash that disrupted air traffic in Europe.

Iceland, too, lies on a hotspot. The island is part of the Mid-Atlantic Ridge, the crack running down the middle of the Atlantic Ocean where the Eurasian and American plates are moving apart. The open crack is filled up with basaltic magma from the upper mantle. Such mid-ocean ridges occur in all oceans, and form a range of undersea volcanoes more than 40,000 kilometres long. At Iceland, however, the Mid-Atlantic Ridge emerges above water. At Þingvellir (Thingvellir) you can walk from the American plate through the gorge of the spreading ridge to the Eurasian plate, a very impressive experience.

The reason the Mid-Atlantic ridge rises above the water at Iceland is because of the hotspot that lies below it: it produces so much volcanic material that Iceland has grown above sea level. The Icelandic hotspot may even be the reason that Europe and America started to drift apart here 55 million years ago. It has left a trail of extinguished volcanoes on both the American and the European plates. Examples are the basalt rocks of the famous Giant's Causeway in Northern Ireland and the remains of volcanic activity in western Scotland.

The hotspot produces a fascinating play of ice and fire. Iceland has the fire of the hotspot to thank for its ice caps, but volcanic activity melts

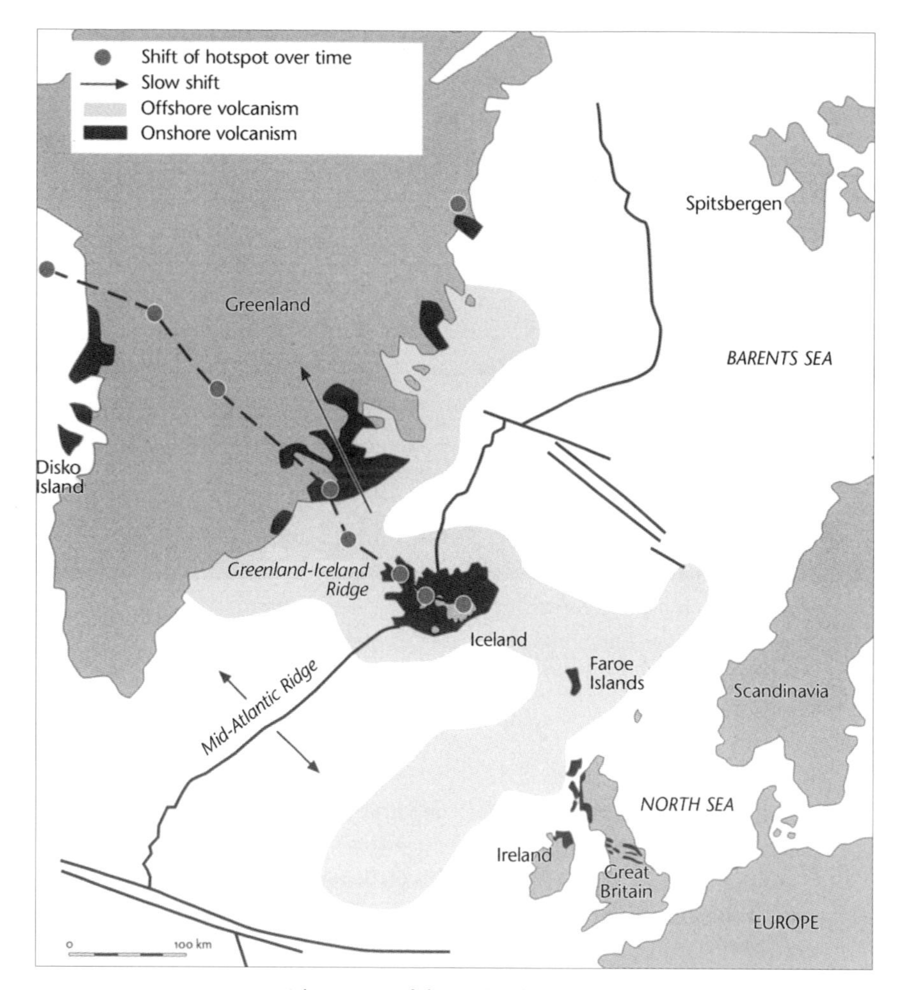

The traces of the Icelandic hotspot.

the ice again and causes *jökulhlaups*, catastrophic meltwater flows, as with the recent eruptions of the Grimsvotn under the ice cap of the Vatnajökull. Some volcanic table mountains were created in the ice age during eruptions that were not strong enough to break through the much thicker ice, a little like a bruise under the skin.

To see exactly where the hotspots come from, we need to use the 'binoculars' of Andrija Mohorovičić again: seismology. Today the paths of so many earthquake waves are known that you can combine them to create a kind of three-dimensional scan of the Earth, just as you can see an unborn baby in its mother's womb. That is known as seismic tomography. This technique has been applied to make the hotspot under Iceland visible: it comes from at least 600 kilometres deep, on the border between the outer and inner mantle; some say even from 2,900 kilometres deep, the border between the mantle and the core.

These are not random borderlines. The propagation velocity of earthquake waves is closely related to the density of the rocks through which they pass: the denser the rocks, the faster they travel. Liquid layers slow them down. A sudden change in velocity signifies a change in the density of the rock. Such changes occur in the mantle at depths of 410 and 660 kilometres. High-pressure laboratory experiments have now shown that, at pressures equivalent to those at a depth of 410 kilometres, olivine – the main component of the upper mantle – transforms into another mineral with a denser structure, wadsleyite. At 660 kilometres other even denser minerals occur, such as ferropericlase. On the surface these minerals are found in places where meteorites have impacted, which also generates extremely high pressures for short periods. The transitions are therefore probably the causes of the sudden changes in propagation speed. On the basis of these transitions, the mantle is divided into the asthenosphere, outer mantle and inner mantle.

At a depth of 2,900 kilometres lies the contact zone between the solid inner mantle and the liquid nickel-iron core: a very active boundary plane that may generate the convection currents that cause plate tectonics, and which is probably the source of many hotspots. Hotspots like Hawaii do indeed seem to come from very great depths, while others lie less deep below the surface. With all these deep magma hotspots, you suddenly see the resemblance to Athanasius Kircher's cross-section of the Earth. In their beautifully illustrated but very technical book *Hoe werkt de aarde?*, Dutch geologists Rob de Meijer and Wim van Westrenen suggest that there may be processes resembling nuclear reactors active in this boundary plane.

Even from such great depths, the Earth has generously sent materials to the surface that we would never have been able to dig up ourselves: diamonds. The discovery of diamonds in South Africa is a hilarious story. Around December 1866, on a farm near Hopetown, a certain Schalk Van Niekerk saw children playing *klip klip* with five small stones. One of the stones looked so different from all the others that he thought that it was a diamond. Everyone laughed at him, until geologist William Guybon Atherstone confirmed that it was indeed a diamond. The stone was sent to London for further investigation, and its authenticity was again confirmed. No one believed, however, that it came from South Africa. 'No one would believe that

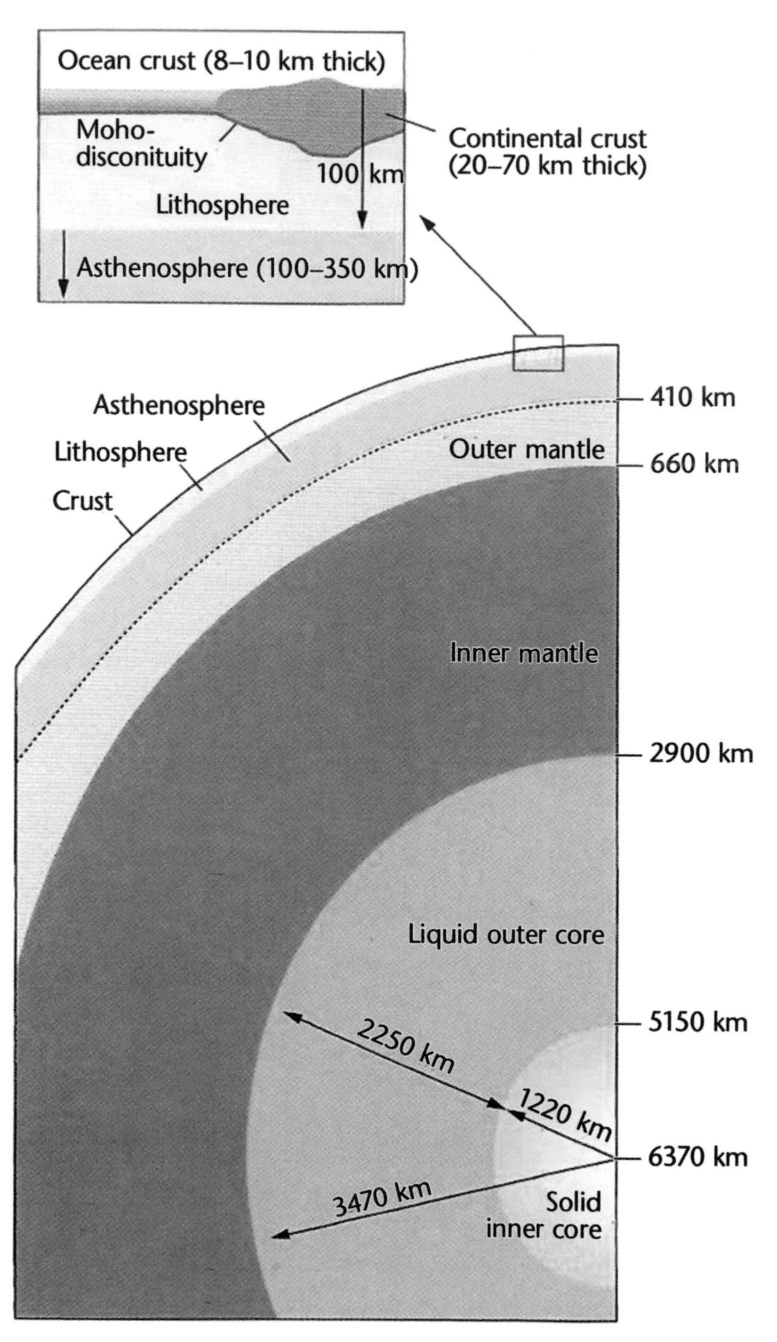

Discontinuities in the Earth.

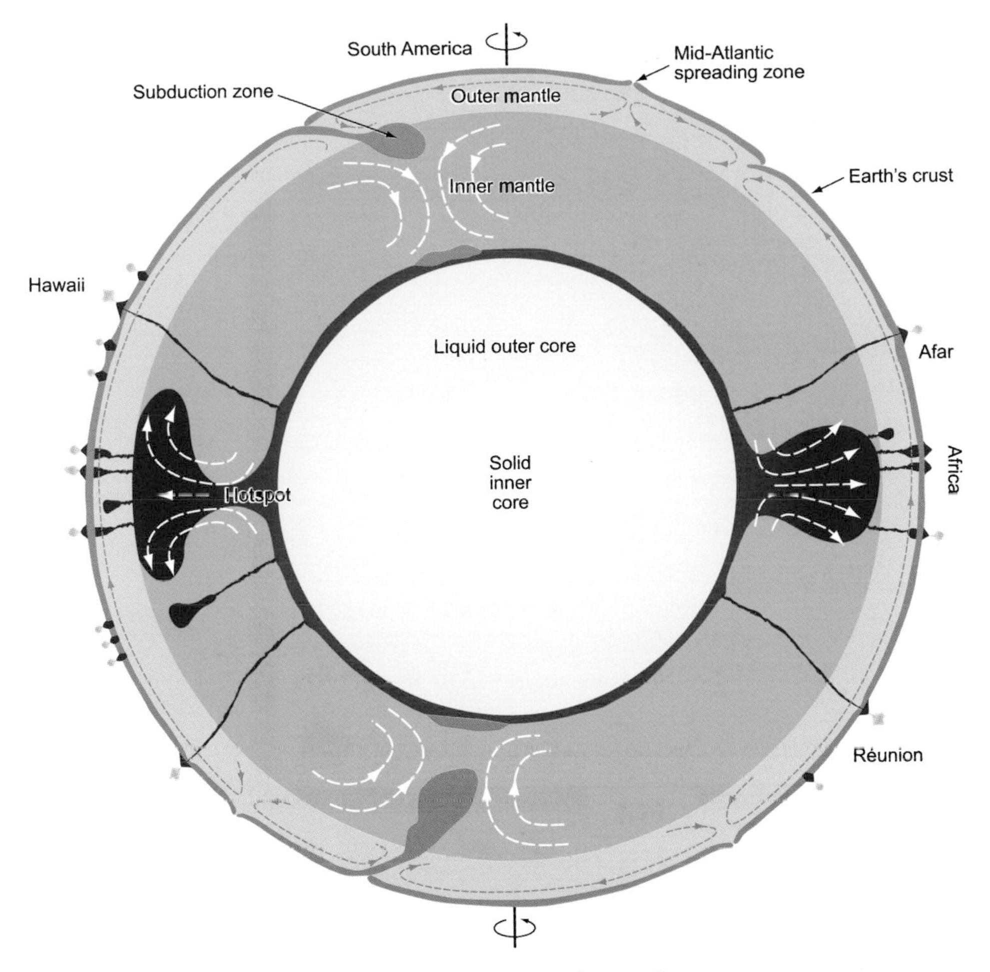

The origin of hotspots in the mantle.

the impoverished Cape could produce such riches', writes Brian Roberts in his book *Kimberley, Turbulent City*.

It was therefore two years before a mineralogist, James R. Gregory, went to South Africa to investigate. He saw as soon as he arrived that there could be no diamonds in the area and that if one had been found, it must have been brought by an ostrich. 'I made a very careful and lengthened examination of the district where the diamonds were said to have been found', he wrote later in *Geological Magazine*, 'but saw no indications whatever that would warrant the expectation of the finding of diamonds, or of diamond-bearing deposits, at any of the localities.' He concluded that the geology of the area was not suitable. He had most likely been sent with orders not to allow the market to be

The first diamond prospecting at Colesberg Kopje, South Africa, 1871–2.

disrupted by new finds. In 1869, however, Van Niekerk reappeared with an 83.5-carat diamond, and the argument was settled.

In July 1871 a man with a beard and wearing a red cap returned agitated from the *veld* claiming that he had found yellow ground and had panned three diamonds out of it. The man, Fleetwood Rawstorne, pointed towards a hill near a small clump of thorn bushes which he called Colesberg Kopje, after the place he was from. That hillock is now a large pit more than 1 kilometre deep: the Big Hole of the Kimberley Mine. As this was the first time that diamonds had been found in source rock, it was named kimberlite, or blue ground. The yellow ground is produced by weathering of the blue ground.

The only possible mitigating circumstance for Gregory's nonchalance is that diamonds often occur in narrow, vertical pipes that take up little space on the surface. In Suriname they have been searching in vain for 30 years for the source of the diamonds in the Rosebel district. Kimberlite consists of chunks of peridotite-like rocks of varying sizes. Both the shape of the pipe and the nature of the rock suggest that it is the supply channel of a volcanic eruption, most probably from a hotspot. No one has ever experienced such an eruption, the youngest pipe in the world being 50 million years old. Because of the pressure required to create diamonds, they have to come from at least 150 kilometres deep in the Earth's mantle. In the Big Hole we encountered them no further than 1 kilometre down from the Earth's surface.

While they are forming in the mantle, diamonds themselves often enclose minuscule crystals of other minerals, in the same way that a twig floating in a stream will be enclosed by ice when the temperature falls

Colesberg Kopje a year later. The scene is reminiscent of the wild gold prospecting of the 1990s in the Serra Pelada, in the Serra dos Carajás in Brazil's Amazon region.

Colesberg Kopje, now the Big Hole, South Africa. The mine closed in 1914.

below freezing. Some of these minerals, such as ferropericlase – already mentioned above – are only stable at pressures higher than those found at 660 kilometres below the surface. This fact makes these diamonds the deepest samples we have from the Earth's interior. The Earth has been unable to gargle up anything for us from any deeper. That is less than one-tenth of the radius of the Earth. The rest is still untouched. This shows how little we really know.

Diamonds are simply carbon. The chemist Humphry Davy, inventor of the miner's lamp, succeeded in igniting a diamond using a burning glass. Carbon is found not only in the atmosphere as CO_2, as bicarbonate in the oceans, as carbonate in limestone, as graphite in schists, as organic molecules in the soil, in oil, in coal and in living organisms, but also in diamonds hundreds of kilometres deep down in the Earth's mantle. Where does that deepest carbon come from? Is it also a product of life? It is not impossible.

Carbon has various isotopes: lighter carbon-12 and heavier carbon-13. The radioactive isotope carbon-14 that we use to date peat, wood and shells has no part in this story. Plants prefer carbon-12 to carbon-13 for photosynthesis, so material originating from plants has more of the lighter isotope than there is in the CO_2 in the air. Some diamonds have so much more carbon-12 than carbon-13 that they, too, may have organic origins.

Images from seismic tomography have shown that subduction can force a lithospheric plate as far as 700 kilometres into the mantle, and perhaps even further. If there are deposits containing organic vegetable matter, such as coal, on the edge of the plate, they will be carried far into the depths. That carbon can be released into the air again during volcanic eruptions in the form of CO_2, or it may be compressed to form graphite, or eventually even diamond. The oldest piece of stone ever found in the world, a zircon crystal 4,252 billion years old in the Jack Hills in Australia, contains a minuscule diamond that has so much of the light isotope carbon-12 that it may even be a sign of life. But that is not certain, as meteorites sometimes contain carbon that is rich in the light carbon isotope.

So if you are cremated, there's no need to ask the funeral director to get your ashes compressed into a diamond: nature can do it for you. But you do need to be patient: it might take 4 billion years for it to happen.

Sulfur, too, plays a role in the depths of the Earth. It combines with iron, nickel or other metals to form sulfides like pyrite, or ferrous

Sulfur crust on the Brennisteinsalda ('sulfur wave') volcano, Iceland.

sulfide. Many mantle and igneous rocks from the deeper Earth's crust contain small sulfide crystals. There may be magmas deep in the crust that consist completely of sulfides, a kind of hot, yellow mercury. Such sulfidic magmas do not combine with silicate magmas like basalt magma, in the same way that oil and water do not mix, for example. If such magma drops concentrate into greater units, sulfidic ores can be created. As we have seen, other sulfidic ores are deposited by hot, watery solutions around black smokers. Water and fire therefore both play a role.

In this way, sulfur that is expelled by volcanoes, such as on Iceland and in the Solfatara, can come from various sources. Magma contains sulfur-bearing gases in dissolved form which separate only when it comes close to the Earth's surface and the pressure decreases, just like carbon dioxide gas in a bottle of carbonated water or beer. Some of that sulfur probably originates from deep within the Earth. However, seawater also contains sulfur in the form of sulfate. When seawater evaporates, gypsum is deposited and, if that ends up deep under the surface of the Earth as a result of subduction, sulfur is also transported to the mantle. That comes back out again later during eruptions as sulfur, sulfur dioxide (SO_2) or hydrogen sulfide (H_2S). Sulfur also has various isotopes but, like those in carbon, they do not always prove conclusively whether it comes primarily from the interior of the Earth or has been recycled from seawater.

The dream of drilling through to the mantle lives on. The Earth's crust is thinnest under the oceans, around 8 kilometres thick. Drilling there would offer the greatest chance of success. In his autobiography Austrian-American oceanographer Walter Munk describes a meeting of the American National Science Foundation in 1957, where they had to select a number of projects for funding from 65 proposals. At one point, the question was asked: 'If cost were no object, what would be the project that could lead to the greatest advance in understanding the Earth?' 'Drilling through to the mantle', replied Harry Hess and Walter Munk himself. They got the money. 'Thank God we're finally talking about something besides space', seemed to be the general response.

The project was called Mohole, a hole to the Moho. In 1961 scientists drilled a test hole close to the island of Guadalupe, off the coast of Baja California, Mexico, through 190 metres of deep sea sediment, and then struck basalt. The test project had cost $1.7 million. But when it came to contracting out the drilling of the actual hole, it all went wrong: the National Science Foundation took over the project from the scientists and hired an engineering company with no experience at all of drilling at sea. There was disagreement about whether they should drill one, two or several holes. Then Mohole became Slowhole. And in 1966, after another $50 million had been spent achieving nothing at all, it became Nohole. In Dante's *Inferno* hypocrites are punished with a gilded lead cloak. That makes them so slow that even fire cannot get them moving.

The next attempt did not take place until 2005, from a Japanese ship, the *Chikyu*. They have now reached 1,400 metres, but currently seem to be more interested in the extraordinary life forms around black smokers on the ocean floor than in the mantle. Immediately following the earthquake of 11 March 2011 they took 200 people on board. In the tsunami that followed the quake, the ship spun around three times, but they survived. 'Our objectives are still there, somewhere in a distant haze', they say apologetically on their website.

FIFTEEN

To the Centre of the Earth

To talk about the bottom of the universe
the way it truly is, is no child's play.
Dante Alighieri, *Inferno*, XXXII:7–8

In Jules Verne's famous book from 1864, Axel *was* allowed to accompany his uncle, the geologist Professor Lidenbrock, in his journey to the centre of the Earth: another boy, another uncle. I have always been jealous of Axel. My copy of the Dutch translation, published by Elsevier in 1912 and beautifully bound in blue linen with gold-leaf letters, was rather dog-eared. I devoured it just as I did the books of Karl May, Verne's German contemporary and kindred spirit, who also wrote exciting stories about places he had never been to. I wanted to go to them, too.

Professor Lidenbrock was an irritable and conceited scholar who gave a not particularly popular course on mineralogy at the Johanneum in Hamburg, 'for himself and not for others' and 'with his mallet, his steel spike, his magnetic needle, his blowlamp, and his flask of nitric acid' he could identify all 600 kinds of minerals. In his study, there were 'specimens of the whole mineral order . . . labelled in the most perfect order', and young Axel much preferred to spend his time learning about them than playing with other boys of his age: something I could certainly identify with.

After finding a parchment with instructions, they travelled to Iceland where, in the Sneffels Joculis (Snæfellsjökull) volcano, the entrance to the interior of the Earth was reputed to lie. I have since followed in their tracks, not with my uncle but with my partner, and not to the end of their journey. The volcano really exists, and is on the west coast of Iceland. Verne writes that the Sneffels can be seen from Reykjavík, 'a high mountain with two points on top, a double cone covered with perpetual snows'. And he was right! But Verne had never been to Iceland, so how could he have known what it looked like?

(top) Eugène Robert, *Snæfellsjökull at Midnight*, 1840; (bottom) Eugène Robert, *The Coast at Stapi*, engraving.

He reveals, rather surreptitiously, how he came by the information. He has an Icelandic scholar, a Mr Fridriksson, tell Lidenbrock that there have been recent scientific reports from Messrs Gaimard and Robert on board the French corvette *La Recherche* and from other French scholars on the frigate *La Reine Hortense*. Verne does not fabricate all this; these voyages really took place and Verne made extensive and detailed use of their reports. Even his informant, Mr Fridriksson, really existed. The first voyage on *La Recherche* in 1835 and 1836 was led by Paul Gaimard and geologist Eugène Robert and was looking for the *La Lilloise*, a ship that had disappeared without trace in

Édouard Riou, 'The Coast at Stapi with Basalt Columns', from Jules Verne's *A Journey to the Centre of the Earth* (1864), engraving.

1833 between Iceland and Greenland. The second expedition in 1857, on *La Reine Hortense*, was under the leadership of the Polish-French explorer Charles-Edmund Chojecki. The ship carrying the supplies on this expedition was called *Le Cocyte* . . .

On board *La Recherche* Eugène Robert made beautiful drawings of the rock formations they encountered en route, which have been included in an atlas that is part of the report on the voyage. Verne studied the drawings closely, as is obvious from his description of the rocks along the coast at Stapi (now Arnastapi) at the foot of Snæfellsjökull.

> The walls of the fjord, like the whole coast of the peninsula, were made up of a series of vertical columns, thirty feet tall. These straight shafts of perfect proportions supported an archivolt made of horizontal columns, whose overhang produced a half-vault over the sea. At intervals in this natural impluvium, one's

eye detected arched openings of a superb design, through which rushed and foamed the waves from the open sea.

The splendid engravings in Verne's *Journey to the Centre of the Earth* by artist Édouard Riou are also inspired by Robert. But Riou was apparently not satisfied with the lack of depth in Robert's pictures and added a few columns himself. He himself had never been to Iceland and clearly interpreted the basalt columns in Verne's text in a more architectural sense. He was awarded the Légion d'Honneur for his illustrations.

This all shows the lengths to which Verne was prepared to go to provide a reliable picture of Iceland. Of course, in the eyes of the reader, it also increases the credibility of what follows: the descent into the Sneffels. The climb up the mountain seems to be accurate, too: a recent lava flow on the southeast side is the most accessible route. The only surprising thing is that, while he first describes the mountain as snow-covered – and Robert also draws it covered with snow on all sides – there is no mention of snow at all during the description of the climb. But there are other careless errors in the text: dates do not add up (1 July 1863 comes before 30 June), and the last eruption of the Sneffels was first reported as 1219 and then later as 1229 (there have been no reports of the volcano erupting in recorded history).

The coast at Arnastapi in 2010.

French geographer Lionel Dupuy, who devoted his thesis to the subterranean journey, discovered that Verne even made an enormous error in calculating the distances below ground.

They reach the top and descend from there into the depths. And even then, the parallels with the real landscape continue to some extent. At the foot of the volcano, on the southeast side, there are at least six lava tunnels: caves created when the outside of a lava flow solidifies while the lava on the inside is still liquid enough to flow out, leaving an empty crust behind. The longest lava tunnel in Snæfellsjökull, Langiþröngur, penetrates at least a kilometre into the mountain. Robert even drew one of them.

Lava tunnels are not unusual, and are sometimes used as ice cellars: the winds are often so strong that they cause severe cooling, and evaporating water turns to ice. The Trou à Glace near Pontgibaud in Auvergne, France, contains ice even in the middle of the summer. The Vatnshellir lava tunnel in Snæfellsjökull is a kind of dripstone cave, with icicles as stalactites. It is tempting to conclude that Jules Verne chose the Sneffels for the descent specifically for that reason.

The descent was not difficult at first. 'We were able to simply let ourselves go on these inclined slopes, without straining ourselves. It was Virgil's *facilis descensus Averni*', says Axel, the narrator. They were

Vertical and horizontal basalt columns, Snæfellsnes, Iceland.

The ascent along a young lava flow on the southeast side of Snæfellsjökull.

Lava tunnel drawn by Eugène Robert.

Stalagmites of ice in the Vatnshellir lava tunnel. 'But what acted as steps under our feet became stalactites on certain walls. The lava, porous in places, was covered with little round bulbs; crystals of opaque quartz, decorated with clear drops of glass, hung from the vaulted ceiling like chandeliers, and seemed to light up as we passed. It was as if the spirits of the underground were lighting up their palace to welcome their guests from the Earth.'

soon 10,000 feet below sea level, 'six thousand feet further than the greatest depths achieved by man, such as the mines of Kitzbühel in the Tyrol or those of Wuttemberg in Bohemia.'

On their way down, they first take the wrong tunnel. They expect to find the granite of the basement immediately under the lava, but instead come across Silurian layers containing trilobites, fossils resembling woodlice that do indeed date from that era. Then they came to the Devonian layer: 'The electric lamp produced a wonderful sparkling on the schists, the limestone, and the Old Red Sandstone of the walls.' They compare it to Devon and again, Verne provides a reliable-sounding list of rocks you would indeed find in that county. Eventually they come to a coal layer from the Carboniferous era. This means they are moving ahead in time, from older to younger layers rather than going deeper, suggesting that Axel's fears that they were going the wrong way were well-founded.

Édouard Riou, inside the 'hollowed-out diamond', from Verne's *A Journey to the Centre of the Earth*, engraving.

But Axel's greatest fear is that it will be far too hot in the interior of the Earth. The presence of so much coal only fuels his fears, since the heat that was required for such luxuriant vegetation can only come from the deepest interior. At the start of the book, a discussion arises between Axel and his uncle: Axel thinks that it becomes a degree warmer with every 70 feet that they descend, which will mean that it should be 2 million degrees in the centre of the Earth. 'The substances at the Earth's core exist therefore as white-hot gases', he says. We hear echoes of Descartes.

But Lidenbrock says that, if this were the case, 'the white-hot gases produced by the fusion of the solids would acquire such force that the Earth's crust could not resist and would explode like the walls of a boiler under steam pressure.' The centre of the Earth cannot be liquid, Lidenbrock continues, otherwise there would be internal tides twice

Gypsum crystals many metres long, Naica cave, Mexico.

a day which would cause periodical earthquakes. Lidenbrock appeals to the theories of Poisson and Davy, whose books most probably also lay on Verne's writing desk. Lidenbrock had an answer to everything. He was made of the same stuff as Karl Ernst von Baer and Athanasius Kircher. But that does not temper Axel's fears. During the descent, he keeps a close eye on the temperature.

They soon find the right passage, where the basement consisted of 'schists, gneisses, and mica-schists resting on that immovable rock called granite'. In many parts of Europe this is indeed the oldest rock. And then: 'The light from the lamps, reflected by the tiny facets of the mass of rock, shone its fiery flashes at all angles, and I imagined I was travelling through a hollowed-out diamond with the rays disintegrating into a thousand dazzling lights.' This wonderful sentence sends your imagination wild, and that's exactly what it did to Édouard Riou. His engraving of that scene is for me the most splendid in the whole book, despite all the other spectacular drawings that follow it.

It is not very probable that Riou had a real example to copy from, but in the year 2000, one was discovered, you might say posthumously, when a wall was broken through in a lead-zinc mine in the Sierra de

Naica in Mexico. The miners found a cavity with dozens of sparkling gypsum crystals up to 7 metres long.

The source of Verne's inspiration may, however, be much more banal, according to the commentary by the English translator of the book, William Butcher (1992): 'There are also disturbing similarities with George Sand's *Laura: Voyage dans le cristal* (1864).' Butcher could be right. In that book, Alexis, a young German mineralogist, dreams that he travels to the hole in the Earth at the North Pole with his sinister Uncle Nasias. The interior of the Earth bears a striking resemblance to a geode, a hollow stone gobstopper lined with crystals on the inside. First they come to an ice-free sea, from which they can see an inaccessible rock face 2 to 3 miles high that, at closer inspection, proves to consist of gigantic perpendicular crystals of tourmaline, a columnar, glossy black mineral. They also see a glacier-covered mountain top that turns out to be made of a single olivine crystal. Then Alexis awakes from his dream.

Verne's adventurers do not descend vertically, but down a gentle slope that continually heads in a southeasterly direction. Then they follow a deep slit, which they conclude was created during contraction

Geode of quartz with calcite crystals in its core, a metaphor for the hollow Earth.

of the Earth at a time when it was cooling down. This was a popular theory of the origin of mountains at the time. On 18 July they arrive at a depth of sixteen leagues, 64 kilometres from the start. Axel shouts out that they should be at the underside of the Earth's crust! This is not a bad estimate, though a little on the thick side, especially since they should now be under the Atlantic Ocean, where the crust is only 5 to 10 kilometres deep. Under high mountains, this could have been possible, as we have seen. Furthermore the ocean crust should be made of basalt, not of granite, but they did not know that in 1864, when Verne wrote the book. Dutch geologist J.H.F. Umbgrove even posited in his famous book *The Pulse of the Earth* in 1942 that there was a thin crust of sial (granitic rock) in the Atlantic Ocean. This idea remained intact until the 1960s. There is still much that we do not know about the Earth, but we have very resolutely managed to forget just how much we did not know in the past.

At that point, Axel starts to feel uneasy, since at that pace they will need to keep walking for five years before they reach the centre of the Earth. According to him, the temperature should have been 1,500°C, while the thermometer stood at 27°C. But Lidenbrock pushes on unperturbed. At 350 leagues, 1,400 kilometres to the southeast of Iceland, they are surprised to come across an immense subterranean sea which Lidenbrock, not one to suffer from false modesty, calls the Lidenbrock Sea. They are now at a depth of 35 leagues, or 140 kilometres. An interesting distance, as this is exactly the depth of the asthenosphere, the viscous layer of the Earth's mantle that plays a crucial role in plate tectonic movements. The asthenosphere was discovered by geophysicist Beno Gutenberg in 1960 but this, too, Verne had foreseen.

According to our travellers, the Lidenbrock Sea is much larger than the Mammoth Cave in Kentucky – now the largest known karst cave in the world, with 630 kilometres of subterranean rivers – and the Cueva del Guácharo in Venezuela (Verne says Columbia, as Humboldt described the cave when the country was still part of Gran Colombia; there is also a Cueva de los Guácharos in present-day Colombia, but that was not discovered until much later). They construct a raft of half petrified tree trunks and set sail until they eventually come to a geyser with a temperature of 165°C, which Axel takes to be proof that the centre of the Earth is indeed hot.

The subterranean Lidenbrock Sea is by no means devoid of life. Blind fish swim in its waters, gigantic mushrooms grow on its shores, and the explorers are almost dragged down in a titanic battle between an ichthyosaurus and a plesiosaurus, monsters from the Jurassic era that have been long extinct on the surface of the Earth, but which have clearly survived here. Later, they also find whole fields of bones of other extinct creatures from the Tertiary era: 'A thousand Cuviers would not have been enough to reconstruct the skeletons of all the once living creatures which now rested in that magnificent bone-graveyard', Verne writes.

Our travellers even stumble across the preserved body of a primitive human and believe they see a living giant: 'There, less than a quarter of a mile away, leaning against the trunk of an enormous kauri tree, was a human being, a Proteus of these underground realms, a new son of Neptune, shepherding that uncountable drove of mastodons!' Perhaps it was the giant Bárður from Stapi, which according to Icelandic sagas disappeared into the glaciers of Snæfellsjökull. They make a quick escape.

They continue their underground journey through all the geological ages. In his thesis Lionel Dupuy concludes that the book is not only a journey in space, but also in time. I like the sound of that, a journey through space *and* time.

The Lidenbrock Sea involuntarily brings to mind the Niflheim, the Underworld of Nordic mythology. And the old tales of Plato, Leonardo and Descartes about the water layer in the Earth, and the hollow Earth

Modern life on the summit of Snæfellsjökull. The postcard's caption reads: 'The Top of the Centre of the Earth'.

of Edmond Halley, Casanova and John Cleves Symmes. Axel remembers the theory of 'a British captain's which compared the Earth to a vast, hollow sphere'.

When they try to continue on their way on the other side of the sea, their path is blocked by a large rock. They succeed in blowing it up, but only then does it become clear what it was there for: it was a plug in a ponor. The whole Lidenbrock Sea empties out and they are carried along by the water until finally, against their will, they are spewed out by the volcano Stromboli. They are back on the surface of the Earth. They did not make it to the centre of the Earth and the question of whether the interior is hot remains unanswered.

Gustave Doré, Lucifer trapped in the ice of the Cocytus, 1857.

SIXTEEN

With Lucifer in the Ice

The king of the vast kingdom of all grief
stuck out with half his chest above the ice;
. . .
not feathered wings but rather like the ones
a bat would have. He flapped them constantly,
keeping three winds continuously in motion
to lock Cocytus eternally in ice!
Dante Alighieri, *Inferno*, XXXIV:28–9, 49–52

Scientists, mining engineers and oil companies have not yet succeeded in penetrating further than 12 kilometres into the interior of the Earth. But even more surprisingly, few writers, poets and science fiction authors have also succeeded in reaching the Earth's core. Jules Verne's *Journey to the Centre of the Earth* is broken off prematurely, Casanova's heroes got no further than Verne's, and even the fantasy of Edgar Burroughs extended no further than Edmond Halley's hollow Earth. Only Dante's imagination was big enough to reach the centre.

But the core of Dante's Earth looks very little like ours. In the deepest part of the Inferno, where we would expect fire, or at least heat, Lucifer sits imprisoned in the ice of the Cocytus. His beating wings create an icy wind that keeps the lake frozen.

In the Bible, hell is a pool of fire, the Islamic hell of the *Mi'raj* is full of fire and, although Verne's Lidenbrock was still not convinced, the core of the Earth according to modern science is also hot: between 4,000 and 7,000°C. Dante undoubtedly knew that it becomes hotter as you penetrate more deeply into a mine. Perhaps he wanted the deceivers and traitors frozen into the plain of ice to see their reflections and thereby their true selves.

Newton had calculated the difference in density between the Moon and the Earth in 1687, but it was not until 1798 that the British physicist Henry Cavendish determined the absolute density of the Earth: 5.48

The real Cocytus, a quiet arm of the Acheron in northern Greece.

times that of water, which is close to the currently accepted value of 5.53 grams per cubic centimetre. And it took another century before the East-Prussian scholar Emil Wiechert deduced in 1896 that the interior of the Earth must have a greater density than on the surface, since normal rocks like granite only had a density of 2.7. Wiechert knew that some meteorites consisted of nickel iron. Many are fragments of exploded planets, leading him to suggest that the Earth also had a core of nickel iron. His calculations supported that argument, but at the time he had no evidence to back it up. Modern researchers believe that the Earth's core is a little too light to consist only of iron and nickel, and there is one theory that it may contain 10 per cent sulfur.

It is astounding that it took another century before someone came up with the idea of testing that experimentally, since as you know that the density is 5.5, you would think it logical to weigh a stone with a similar specific weight. But perhaps the barrier between physics and geology was still too high. Geology did not finally develop as an independent science until the end of the eighteenth and the beginning of the nineteenth centuries, with the work of James Hutton and especially Charles Lyell. And at that time, it was based largely on field observations, something that physicists like Newton and Cavendish were perhaps less likely to indulge in.

British seismologist Richard Dixon Oldham showed in 1906, on the basis of the propagation behaviour of earthquake waves, that the Earth does indeed have a core. Oldham analysed the arrival times of both types of waves from major earthquakes in Japan, Alaska, Argentina, Central Asia and Indonesia at seismographs all over the world, and used them to calculate the speed at which they moved through the Earth. The propagation velocity of the shock waves not only increased as they went deeper, which is normal when pressure and density increase, but almost made strange leaps.

At a depth of six-tenths of the Earth's radius they slowed down rather than speeding up. That was the basis of the argument that the Earth must have a core of material with completely different properties than the rest of the planet: it was the first concrete evidence that the Earth is indeed a gobstopper. 'I do not propose to enter into speculative grounds, or to offer any opinion as to whether this central core is composed of iron, surrounded by a stony shell, or whether it is the central gaseous nucleus of others', Oldham writes. Descartes' ideas about a sun-like substance in the interior of the Earth were still alive and kicking a century ago.

But that theory was soon disproved. S-waves cannot propagate through liquid and, since they also did not penetrate the core of the Earth, Emil Wiechert's pupil Beno Gutenberg concluded that the core must be made of liquid nickel iron and that this was the source of the Earth's magnetism. Later, together with Charles Richter, Gutenberg devised the Gutenberg-Richter scale for measuring the magnitude of earthquakes and, even later, discovered the asthenosphere.

The Greek writer Plutarch, who lived in the Roman era (around CE 46–120) knew about the liquid nickel-iron core of the Earth 2,000 years ago: in the *Vision of Aridaeus*, his hero Thespesius descends into hell. Sinners who have to pay for their envy and ambition are first submerged in a bath of boiling gold. When they start to glow and become transparent, demons fish them out with large tongs and throw them into a second bath of frozen lead. When the souls are frozen and as hard as hailstones, the demons throw them into a bath of molten iron. They become so horrifically black and stiff that they split apart. Then they have to go back into the first bath and the cycle goes on endlessly. It was like Dante *avant la lettre*.

In 1936 Danish seismologist Inge Lehmann wrote an article with, to my knowledge, the shortest title of any scientific publication ever: 'P'' (P-prime). At the time, that was the sign used to designate a P-wave that

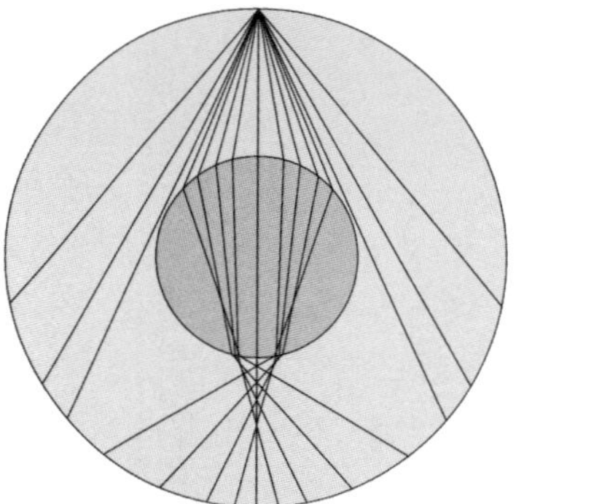

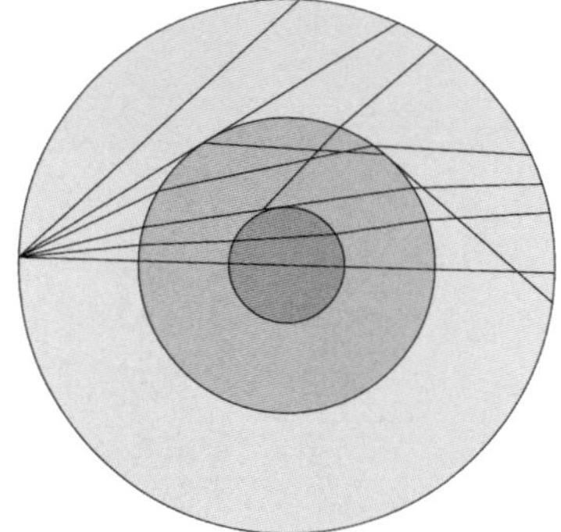

(left) Original figure showing Oldham's core from 1906. (right) Inge Lehmann's solid core from 1936.

had passed through the Earth's core before being detected by a seismograph. In 1929, to her surprise, Lehmann had picked up a signal from an earthquake in New Zealand that should actually have been in the shadow of the core. That led her to conclude that there must be a solid inner core in the liquid outer core. And that brings us to the modern image of the gobstopper.

The only hell Lehmann saw was that of a woman in a man's world: 'You should know how many incompetent men I had to compete with – in vain', she said. I open *Algemeene Geologie* by Berend G. Escher once again. Inge Lehmann's solid inner core is not mentioned – it was too soon. When my uncle started studying biology in Leiden in 1937, he probably heard nothing about it, just as I never heard anything about plate tectonics while I was studying in the 1960s. Lipke must have bought the book when he was in the first year: 'L. B. Holthuis, 1938', he wrote in it; I wrote 'January 1963' next to that, under my name stamp, which I had made in a now incomprehensible fit of bureaucratic fervour. And suddenly it occurs to me that this gift perhaps had more significance than purely to feed my embryonic interest in geology. The fact that he had given the book away speaks volumes for someone whose life, apart from crabs, revolved almost entirely around books. When he died in 2008, we – his nephews and nieces, he had no other family – cleared his private apartment, and we had to sort through all his books. Any with crabs and crayfish in them had to go to the museum, but the rest were for us or the purchaser of his remaining effects. But he no longer had any need of *Algemeene geologie* already back in 1963. He clearly had no more interest in geology. How much

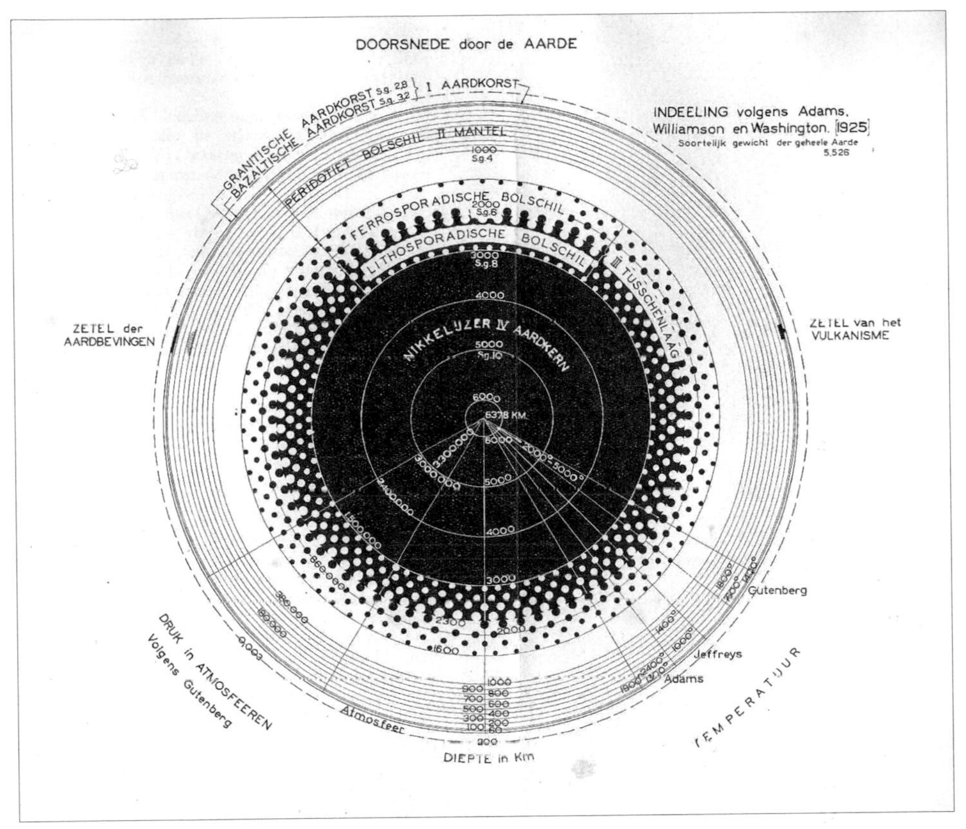

The structure of the Earth's interior, from B. G. Escher's textbook *Algemeene Geologie* (1934).

I wished it had been different. I came across a copy of my own thesis; I couldn't see if he had read it, but he still had it.

So how did the gobstopper actually evolve? Why are the heaviest components of the Earth at its core, and why do they get lighter as you move outwards? Surely that is no coincidence? We largely have Emanuel Swedenborg, the Swedish scholar who described the copper mine at Falun as a temple of Venus, to thank for the ideas we have about that now. Swedenborg proposed in 1734 that the solar system originated in a cloud of gas that gradually coagulated. A flat disc of gas and dust particles circled around the Sun, which gradually clustered together, eventually forming planets. Through a process of separation, the heaviest elements soon concentrated closer to the Sun and formed the rock planets Mercury, Venus, Earth and Mars, while the lighter components were flung outwards and eventually formed the large gas planets Jupiter, Saturn, Uranus and Neptune.

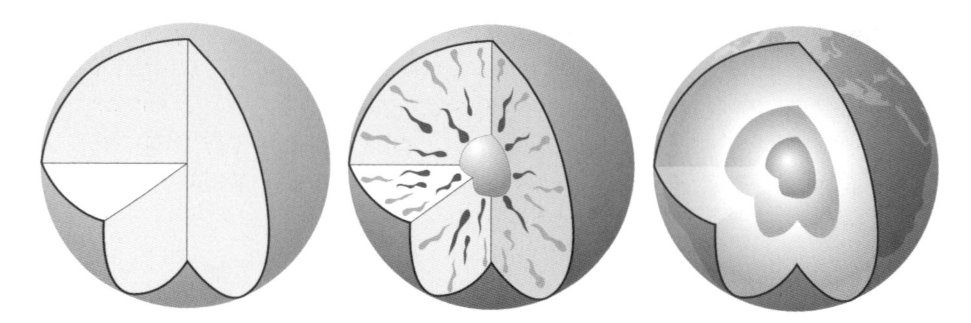

(left) The chondritic, homogeneous Earth. (centre) The sifting process: the heavier components migrate towards the core, the lighter components outwards. (right) The current structure.

At first the composition of a rock planet like Earth must have been homogeneous. Meteorites, known as chondrites, have been found that still have this primitive, undifferentiated form: their composition is the same as the average composition of the Earth. From that homogeneous mixture, the Earth's core, mantle and crust developed, probably because the chondritic material melted as a result of the intense compaction of material under the influence of gravity. The heaviest elements, iron and nickel, sank down to the core; the lighter peridotites accumulated in the mantle, and the even lighter granites in the crust. The water mantle and the atmosphere are the lightest examples of this trend. This was the second round of sifting, after the formation of the planets. It is very similar to what Thomas Burnet described long ago: the heavy particles sank first and the lighter later. This model of course leaves no room for a layer of water inside the Earth, as Plato, Leonardo and Descartes imagined. But we do not, as yet, have samples of rock from the core.

One of the few people to have given serious consideration to how we can penetrate deeper into the interior of the Earth is the renowned American geophysicist David Stevenson. In 2003 he published his idea of sending an unmanned capsule (SubTerraMobile) to the Earth's core. Why is there a NASA for space and not for the underground? he complained. He proposed creating an enormous fissure in the Earth by causing an explosion with the power of several tons of TNT, equal to an earthquake measuring 7 on the Richter scale, and filling the crack with liquid iron. The crack would then propagate itself, moving inwards by melting the rock it passed through. A capsule could be placed in the liquid iron, and would be carried along into the interior of the Earth. The capsule would emit acoustic signals to send information back to the surface. He concludes his argument by saying: 'This proposal is modest compared with the space program, and may seem unrealistic only because little effort has been devoted to it. The time has come for action.' Do it, I say. It hasn't happened yet, but if it ever comes to the first manned descent, I'd advise them to take thick parkas with them, just in case Dante proves to have been right.

SEVENTEEN

The Way Back

Below somewhere there is a space, as far
from Beelzebub as the limit of his tomb,
known not by sight but only by the sound

of a little stream that makes its way down here
through the hollow of a rock that it has worn,
gently winding in gradual descent.

My guide and I entered that hidden road
to make our way back up to the bright world.
Dante Alighieri, *Inferno*, XXXIV:127–35

These lines are about all that Dante has to say about his return from the deepest parts of hell to the surface. That's a shame; a missed opportunity. According to Galileo's calculations, hell is only under 7 per cent of the Earth's surface, below a circle with a radius of more than 1,000 kilometres with Jerusalem at its centre. In other words, it is mostly under Europe and the Middle East. Below the rest of the Earth's crust there is no hell. It would therefore have been a fine tribute to science if Dante had told us what you encounter if you travel to the centre of the Earth without passing through hell.

I've tried it myself. Throughout this book, I followed in the tracks of Dante and Virgil, like a shadow. Now I'm going to follow them back to the surface, but at my own pace, so that I can take the time to write down what I see.

The solid nickel-iron core of the Earth does indeed seem to be crystalline, as some researchers write on the basis of their seismic data. Wonderful: George Sand was right. Just like Alexis, I am walking through the inside of a crystal. In fact I don't even need to walk, as the Earth's core rotates independently, an egg yolk with its own will, a little faster than the rest of the Earth, like a hamster in a wheel. It's nice to experience that. Then I come to the liquid outer core. That is a risky undertaking, as here I have to move through glowing metal, like Thespesius in Plutarch's hell. What's more, the magnetism is very intense here, and is never satisfied with staying in one place. Like a dog scratching around in its basket because it can't get comfortable, it keeps fidgeting around, so that the Earth's magnetic poles continually shift or even reverse their positions. My steel dentures fly out of my mouth.

Then I feel the boundary plane between the outer core and the stone shell, a moment of intense turbulence like in an aeroplane, but I find a place on the head of a rising hotspot and soar like a balloonist up through the mantle.

Everything is fine until I come to around 700 kilometres deep in the mantle. I suddenly notice that I have blisters on my skin and spots, warts and ulcers on my face, and my fingers turn green from gangrene. What is going on? I look on my Geiger counter and see that I have come to an area of extremely high radioactivity. How is that possible? No one predicted that! De Meijer and Van Westrenen thought that the boundary between the core and the mantle at 2,900 kilometres deep could be a kind of nuclear reactor. I didn't notice anything like that there, but here it is a mystery. I quickly don my lead cloak and the symptoms disappear immediately.

But where does the radioactivity come from? Did they do it after all? Years ago it was suggested that you could best dump radioactive waste from nuclear power stations in the 8- to 12-kilometre-deep deep sea trenches along the coasts of the Ring of Fire in the Pacific Ocean. It would then be sucked down into the mantle by subduction and we would be rid of it. There was a lot of protest at the time, especially because of the slowness of the process and the consequences for marine life in the deep oceans. The plan was abandoned, but perhaps they did in secret. It wouldn't surprise me. It is in any case exactly the depth to which we have been able to follow the plate that has been pushed below the surface using Moho's binoculars. It seems to be the first sign of human intervention.

After that it calms down again, and I rise up through the outer mantle. I have to tread water a little so as not to drown in the somewhat more liquid asthenosphere, the Lidenbrock Sea, and cross the Mohorovičić discontinuity until I eventually arrive at the Earth's crust. I am almost home.

But at 12 kilometres, the problems start. I stop rising as smoothly as before: there is something in the way. I observe the obstacle attentively and realize what it is: a diamond! But there is no kimberlite to be seen, and when I study it more closely, I see that it is the remains of the diamond crown of a drill bit. A little later, I see more of them. And then I suddenly see the stinking sludge of dead organic matter, bacteria from the deep biosphere that have been killed by the toxic, barium-rich drilling fluid. I continue to rise but am increasingly pestered by drill-bits. These are not dead, like the one I saw deeper down, but are persistently drilling through the Earth's crust in search of oil. They come towards me, thirsting for the liquid in my body. I can hardly hold them off. They are like the trunks of hundreds of elephants all wanting to suck up the last drops of water from a dried-out lake. At the same time I see enormous colonies of exotic bacteria that are not native to these depths but which spout out of the drill strings – with the exception of those that are in contact with the drill bit and have to transport the oil – blocking up all the pores in the rock. I use all my strength to wriggle loose from the crushing embrace of the writhing strings and reach the top of the oil window. What have I done that they try to punish me so?

But the misery is not over. Suddenly I start to itch terribly. I see small trigonal crystals growing rapidly all over my skin. It looks as though I am developing scales. I scratch them. They are soft: hardness 3. They react with hydrochloric acid. It is effervescent. It's calcite! Calcium carbonate! Limestone! Will I never be rid of this damned limestone? Where is it coming from now? Why didn't Dante say anything about this?

Oh yes, the myth of the twenty-first century, I had almost forgotten it. The sorcerer's apprentices thought that they could influence the climate by filling empty gas fields with CO_2. Carbon dioxide! It occurs naturally deep in the mantle; it is captured in enclosures in my granulite samples from Suriname. There is 40,000 times more CO_2 stored in limestone in the Earth's crust than there is in all the air, sea, soil, plants and animals put together, there is carbon in diamonds, meteorites and comets, and they want to remove a minuscule quantity from the atmosphere because they think it might be harmful. It is the base material of life. Even when the climate started to cool down on its own, they stubbornly hung on to their delusions: they wanted to put all that CO_2 back to warm the atmosphere up again, but that was no longer possible, of course, since it had all been absorbed by the limestone. And now there is another crystallization nucleus in the equation: my body. For a short while, it hardens so quickly that I feel like Lucifer, frozen solid in the ice.

But then I am saved from this fate, too: an ice-cold stream of water suddenly flows past my body, dissolving all the limestone again. The water spouts out of a pipe from above, and a little further away I see another that sucks warm water upwards from the same layer. I know the principle: it is geothermal energy. You hope that they are right when they say that the rapid cooling of the planet caused by this process is negligible. I am relieved: I can continue on my way.

However, my problems are not over. I find myself in a hollow space. It looks like Zipaquirá here, with beautiful salt layers in the rock, but my Geiger counter is going wild again, and I see passages full of rusty barrels piled up all higgledy-piggledy, covered in damp spots and with long beards of salt hanging out of them. Is it salt, or something more sinister? I have to get out of here quickly. I climb on in haste, and immediately have to flee from a tunnel drilling machine that comes towards me threateningly, pulverizing everything in its way like a crazy windmill. There are more and more holes, hollow spaces, mine tunnels, boreholes, cisterns, metro tubes, underground stone quarries, catacombs full of skulls and bones, cave dwellings. I wriggle through hollow layers full of parked cars, the ceilings sagging under the weight of everything on top, and the spaces are increasingly pressed together, like Chloé's house in Boris Vian's *L'écume des jours*, which gets smaller as she gets sicker, because the Earth is sick, all that drilling, hammering and banging is getting more and more deafening, I flee upwards, and become entangled in a jumble of cables and pipelines, I want to get out of here!

And then suddenly, all pieces of the puzzle fit together: this is hell under construction. We are digging and drilling, deeper and deeper, until we have reconstructed that terrible hollow cone from Dante's *Inferno*, with all its vengeful twists and turns, right down to the centre of the Earth. There was no hell; or maybe there was one once, but no one has ever been able to find it. I did my best, but perhaps it lies below the opposite pole to Jerusalem, in the most solitary, most islandless part of the Pacific Ocean. But now the world clearly has urgent need of a real hell. They will show all those unbelievers and sceptics like me that hell really does exist, and on the Day of Judgment all those sinful cities will collapse into the metro tunnels they dug themselves, the oil wells will find the volcanic hotspots they seek and tar and sulfur will spout upwards along with the lava. All the traces of the sinful history of the Earth will be ground to unrecognizable black mire, and the hole in the Earth will be so great that helicopters flying overhead will be sucked down into the depths by the hard maelstrom winds.

And then, suddenly, I am outside. *E quindi uscimmo a riveder le stelle* – and we came out to see the stars once more. At this point, Dante and Virgil were ready to surrender themselves to the astronomers to ascend to Purgatory and Paradise. But this is going too far for me.

I am staying here. I'm going to set up an underground nature reserve. I want to make sure that there is at least one place on Earth, even if it is less than 7 per cent of the planet's surface, where my grandchildren can put on their Turkish slippers and see how the Earth looks in its natural state, with its beautiful fine layers of soil, its podzol, the humble lives of its crawling snails, crabs and beetles, its legacy of dead monsters, its sparkling ore, its plentiful oilfields, its diamonds with their secret traces of ancient life, its slow, deep mantle currents. That is Paradise enough for me.

E quindi uscimmo . . .

Sources

The quotations from Dante's *Divine Comedy* at the beginning of each chapter are from the translation by Mark Musa, published in the Penguin Classics series.

Sources are listed only in the chapters in which they first appear.

TWO: Jerusalem

Abou-Deeb, J. M., M. M. Otaki, D. H. Tarling and A. L. Abdeldayem, 1999, 'A Palaeomagnetic Study of Syrian Volcanic Rocks of Miocene to Holocene Age'. *Geofísica Internacional* (Mexico), 38/1, 17–26

Arkin, Y., and A. Ecker, 2007, *Geotechnical and Hydrogeological Concerns in Developing the Infrastructure around Jerusalem*. Geological Survey of Israel, report GSI/12/2007

Barkay, G., M. J. Lundberg, A. G. Vaughn, B. Zuckerman and K. Zuckerman, 2003, 'The Challenges of Ketef Hinnom: Using Advanced Technologies to Reclaim the Earliest Biblical Texts and Their Context'. *Near Eastern Archaeology*, 66/4, 162–71

—, M. J. Lundberg, A. G. Vaughn and B. Zuckerman, 2004, 'The Amulets from Ketef Hinnom: A New Edition and Evaluation'. *Bulletin of the American Schools of Oriental Research*, 334, 41–71

Bentor, Y. K., 1989, 'Geological Events in the bible'. *Terra Nova*, 1, 326–38

Friedman, T., 1985, 'Quarrying History in Jerusalem'. *New York Times*, 1 December 1985

Hachlili, R., 2005, *Jewish Funerary Customs, Practices and Rites in the Second Temple Period*. E. J. Brill, Leiden

Holy Bible, Authorized Version (King James, 1611)

James, M. R., ed., 1924, 'The Gospel of Nicodemus, or Acts of Pilatus'. In: *The Apocryphal New Testament*, trans. and notes by M. R. James. Clarendon Press, Oxford

Larkin, C., 1920, *Rightly Dividing the Word*. Glenside, PA

Nissenbaum, A., and I. R. Kaplan, 1966, 'Origin of the Be'eri (Israel) Sulfur Deposit'. *Chemical Geology*, 1, 295–316

Singer, A., and A. Ehrlich, 1978, 'Paleolimnology of a late Pleistocene-Holocene Crater Lake from the Golan Heights, Eastern Mediterranean'. *Journal of Sedimentary Research*, 48, 1331–40

THREE: The Wanderings of Odysseus

Anon., n.d., *The Greek Tradition in the Burial Cult of the Bosporus*. Autoreferaat, accessed at pda.coolreferat.com (in Russian)

Baer, K. E. von, 1872, *Beiträge zur Kenntniss des Russischen Reiches und der angrenzenden Länder Asiens*. St Petersburg

—, 1875, 'Homer's Kenntnisse von der Nordküste des Schwarzen Meeres'. *St Petersburger Zeitung*, den 14. (26.) Juli 1875

—, 1878, *Über die Homerischen – Lokalitäten in der Odyssee*. Verlag Friedrich Vieweg & Sohn, Braunschweig

Chepalyga, A., Ya. Izmailov and V. Zin'ko, 2007. *Field Trip Guide igcp 521–481 Joint Meeting and Field Trip*, Gelendzhik, Russia. UNESCO-IUGS-IGCP-INQUA. Rosselkhozakademiya Printing House, Moscow

Herodotos, 2007, *Het verslag van mijn onderzoek. [Historiën]*. Trans. with notes and introduction Hein L. van Dolen. Fourth edition. SUN, Nijmegen and Amsterdam

—, *The History*. Trans. George Rawlinson (n.d.). Accessed at ebooks.adelaide.edu.au.

Homer, *The Iliad*. Trans. Samuel Butler (1898). Accessed at ebooks.adelaide.edu.au

—, *The Odyssey*. Trans. Samuel Butler (1900). Accessed at ebooks.adelaide.edu.au

Lermontov, M., n.d., 'Taman'. In: *A Hero of Our Time*, trans. J. H. Wisdom and M. Murray. Accessed at www.rt.com

Lukina, T. A., ed., 1984, *Scientific Legacy: The Caspian Expeditions of K. E. von Baer in the years 1853–1857. Diaries and Other Documents*. Nauka, Leningrad (in Russian)

Naber, S. A., 1877, 'Gladstone over Homerus'. *De Gids*, 41, 396–416

FOUR: The Entrance to Hell

Abatino, M., 2002, 'Biologia degli uccelli'. In: E. Abatino, ed., *La Solfatara nei Campi Flegrei*, 157–78, IREDA, Naples

Avallone, A., P. Briole, C. Delacourt, A. Zollo and F. Beauducel, 1999, 'Subsidence of Campi Flegrei (Italy) Detected by SAR Interferometry'. *Geophysical Research Letters*, 26/15, 2303–06

Biraschi, A. M., 1994, 'Introduzione'. In: Strabone, *Geografia, L'Italia*. Biblioteca Universale Rizzoli, Milan

Blas de Roblès. J.-M., 2010, *Where Tigers are at Home*. Dedalus, Sawtry, Cambridgeshire

Boccaccio, Giovanni, 1337, *Rime. Parte Prima*, LXI. Accessed at www.boccaccio.scarian.net

Bodnar, R. J., C. Cannatelli, B. De Vivo, A. Lima, H. E. Belkin and A. Milia, 2007, 'Quantitative Model for Magma Degassing and Ground Deformation (Bradyseism) at Campi Flegrei, Italy: Implications for Future Eruptions'. *Geology*, 35/9, 791–4

Braun, G., and F. Hogenberg, 1572, *Civitates Orbis Terrarum*. Cologne

Breislak, S., 1792, *Essais minéralogiques sur la Solfatare de Pouzzole*. Janvier Giaccio, Naples

Briole, P., A. Avallone, F. Beauducel, A. Bonforte, V. Cayol, C. Deplus, C. Delacourt, J.-L. Froger, B. Malengreau and G. Puglisi, 1999, 'Interférométrie radar appliquée aux volcans: cas de l'Etna et des Champs Phlégréens (Italie)'. CNFGG Rapport quadriennial 95–8, 121–8

Castagnoli, F., 1983, 'Commentaires topographiques a l'Éneide'. *Comptes Rendus des séances de l'Académie des inscriptions et belles-lettres*, 127/1, 202–15

Chiodini, R., R. Cioni, M. Guidi, B. Raco and L. Marini, 1998, 'Soil CO_2 Flux Measurements in Volcanic and Geothermal Areas'. *Applied Geochemistry*, 13/5, 543–52

Clark, R. J., 1996, 'The Avernian Sibyl's Cave: From Military Tunnel to Mediaeval Spa'. In: Holger Friis Johansen, ed., *Classica et Mediaevalia, Revue danoise de Philologie et d'Histoire*, 47/9, 1–29. Museum Tusculanum Press, Copenhagen

Conard, N. J., 2009, 'A Female Figurine from the Basal Aurignacian of Hohle Fels Cave in Southwestern Germany'. *Nature*, 459, 248–52

Cool, H., 1906, 'Agnano'. *Elsevier's Geïllustreerd Maandschrift*, January 1906, 321–31

D'Emmanuele di Villa Bianca, R., R. Sorrentino, P. Maffia, V. Mirone, C. Imbimbo, F. Fusco, R. De Palma, L. J. Ignarro and G. Cirino, 2009, 'Hydrogen Sulfide as a Mediator of Human Corpus Cavernosum Smooth-Muscle Relaxation'. *Proceedings of the National Academy of Sciences*, 106, 4513–18

De Vivo, B., and A. Lima, 2006, 'A Hydrothermal Model for Ground Movements (Bradyseism) at Campi Flegrei, Italy'. In: B. De Vivo, ed., *Volcanism in the Campania Plain: Vesuvius, Campi Flegrei and Ignimbrites*, 289–317. Elsevier, Amsterdam

Di Vito, M. A., L. Lirer, G. Mastrolorenzo and G. Rolandi, 1987, 'The 1538 Monte Nuovo Eruption (Campi Flegrei, Italy)'. *Bulletin of Volcanology*, 49, 608–15

—, R. Isaia, G. Orsi, J. Southon, S. de Vita, M. D'Antonio, L. Pappalardo and M. Piochi, 1999, 'Volcanism and deformation since 12,000 years at the Campi Flegrei caldera, Italy'. *Journal of Volcanology and Geothermal Research*, 91, 221–46

D'Oriano, C., E. Poggianti, A. Bertagnini, R. Cioni, P. Landi, M. Polacci and M. Rosi, 2005, 'Changes in Eruptive Style during the AD 1538 Monte Nuovo Eruption (Phlegrean Fields, Italy): The Role of Syn-Eruptive Crystallization'. *Bulletin of Volcanology*, 67, 601–21

Eboli, P. da, 1220, *Trattato dei bagni/De virtutibus balneis Puteolanis et Baiarum*. Manuscript

Escher B. G., 1934, *Algemeene Geologie*. Fourth edition, Wereldbibliotheek, Amsterdam

Ferrari, G. B. de, 1826, *Nuova Guida di Napoli, dei contorni, di Procida, Ischia e Capri*. G. Glass editore, Naples

Findlen, P., 2004, *Athanasius Kircher: The Last Man Who Knew Everything*. Routledge, London

Hamilton, W., 1772, *Campi Phlegraei: Observations on the Volcanos of the Two Sicilies*. Cadell, London. Accessed at www.gutenberg.org.

Heres, T. L., 1998, 'Cumae, een bezoek aan de oudste "Nieuwe Wereld van de Grieken"'. *Hermeneus*, 70/5, 289–97

Ippolito, F., G. Marinelli, 1981, 'Alfred Rittmann'. *Bulletin of Volcanology*, 44/3, 217–21

Kircher, A., 1682, *d'Onder-Aardse Weereld*, trans. of *Mundus subterraneus*, 1664. Van Waasberge, Amsterdam

Lirer, L., P. Petrosino and I. Alberico, 2001, 'Hazard Assessment at Volcanic Fields: The Campi Flegrei Case History'. *Journal of Volcanology and Geothermal Research*, 112, 53–73

Lucretius, *On the Nature of Things*. Trans. H.A.J. Munro (1908). Accessed at www.scribd.com

Lycophron of Chalcis, *Alexandra*. Trans. A. W. Mair (n.d.). Accessed at www.theoi.com.

Lyell, C., 1997, *Principles of Geology*. Penguin Books, London

Maiuri, A., 1934, *I Campi Flegrèi, dal sepolcro di Virigilio all'Antro di Cuma*. Istituto Poligrafico e Zecca dello Stato, Rome

Majo, E., 1927, 'I fenomeni vulcanici della Grotta del Cane (Campi Flegrei) in rapporto alle variazioni atmosferiche'. *Bulletin Volcanologique*, 4/1, 84–92

Martini, M., L. Giannini, A. Buccianti, F. Prati, P. Cellini Legittimo, P. Iozzelli and B. Capaccioni, 1991, '1980–1990: Ten Years of Geochemical Investigation at Phlegrean Fields (Italy)'. *Journal of Volcanology and Geothermal Research*, 48, 161–71

Mastrolorenzo, G., 1994, 'Averno tuff ring in Campi Flegrei (South Italy)'. *Bulletin Volcanologique*, 56, 561–72

Orsi, G., 2004, *The Neapolitan Active Volcanoes (Vesuvio, Campi Flegrei, Ischia): Science and Impact on Human Life*. 32nd International Geological Congress, Field Trip Guide Book B–28, Florence

—, S. De Vita and M. di Vito, 1996, 'The Restless, Resurgent Campi Flegrei Nested Caldera (Italy): Constraints on its Evolution and Configuration'. *Journal of Volcanology and Geothermal Research*, 74, 179–214

Ovid, *Metamorphoses*. Trans. A. S. Kline (n.d.). Accessed at etext.virginia.edu

Plinio, 1982, *Storia Naturale I*, Einaudi Editore, Turin

Rittmann, A., 1950, 'Sintesi geologica dei Campi Flegrei'. *Bollettino della Societa Geologica Italiana*, 69, 117–28

—, 1936, *Vulkane und ihre Tätigkeit*. Ferdinand Enke Verlag, Stuttgart

Rosi, M., and R. Santacroce, 1984, 'Volcanic Hazard Assessment in the Phlegraean Fields: A Contribution Based on Stratigraphic and Historical Data'. *Bulletin volcanologique*, 47/2, 359–70

—, and A. Sbrana, 1987, *Phlegraean Fields*. Quaderni de 'La Ricerca Scientifica' 114, vol. 9, CNR, Rome

Seneca, 2002, *Ricerche sulla natura*. A cura di Piergiorgio Parroni. (*Naturalia Quaestiones*). Mondadori, Milan

Strabone, 1994, *Geografia, L'Italia*. Biblioteca Universale Rizzoli, Milan

Todesco, M., G. Chiodini and G. Macedonio, 2003, 'Monitoring and Modelling Hydrothermal Fluid Emission at La Solfatara (Phlegrean Fields, Italy):

An Interdisciplinary Approach to the Study of Diffuse Degassing'. *Journal of Volcanology and Geothermal Research*, 125, 57–79

Twain, M., 1869, *The Innocents Abroad*. American Publishing Company, Hartford, CT

Valentino, G. M., G. Cortecci, E. Franco and D. Stanzione, 1999, 'Chemical and Isotopic Compositions of Minerals and Waters from the Campi Flegrei Volcanic System, Naples, Italy'. *Journal of Volcanology and Geothermal Research*, 91, 329–44

Varriale, R., 2008, '*La Grotta del Cane: l'esplorazione ed il rilievo di un geosito artificiale ipogeo nell'area vulcanica dei Campi Flegrei*'. Atti VI Convegno Nazionale di Speleologia in Cavità Artificiali. Napoli, 30 maggio–2 giugno 2008 OPERA IPOGEA 1/2, 315–33

Virgil, *The Aeneid*. Trans. A. S. Kline (2002). Accessed at www.poetryintranslation.com

—, *The Georgics*. Trans. A. S. Kline (2002). Accessed at www.poetryintranslation.com

Webster, J. D., M. Sintoni and B. De Vivo, 2006, 'The Role of Sulfur in Promoting Magmatic Degassing and Volcanic Eruption at Mt. Somma-Vesuvius'. In: B. De Vivo, ed., *Volcanism in the Campania Plain: Vesuvius, Campi Flegrei and Ignimbrites*, 219–33, Elsevier, Amsterdam

Welter-Schultes, F. W., and I. Richling, 2000, 'Palaeoenvironmental History of the Holocene Volcanic Crater Lake Lago d'Averno (central southern Italy) Inferred from Aquatic Mollusc Deposits'. *Journal of Quaternary Science*, 15/8, 805–12

Young, T. G., 1878, 'The Gas of the Grotta del Cane'. *Journal of the Chemical Society, Transactions*, 33, 51–2

FIVE: The Vestibule

Apuleius, *The Golden Asse*. Trans. William Adlington (1639). Accessed at ebooks.adelaide.edu.au

Bokhorst, J., 2010 'Graaf vaker een kuil'. In: J. Bouma et al., eds, *Profiel van de Nederlandse Bodemkunde, 75 jaar Nederlandse Bodemkundige Vereniging, 1935–2010*. Nederlandse Bodemkundige Vereniging, 134–7

Dobrovolski, G. V., 2002, '150 Years of Soil Research Resources in Russia'. In: *Russia and the surrounding world*, 5, 1–16 (Russian). Yearbook 2002, Ecological Faculty, Independent International University of Politics and Ecology (MNEPU), Moscow

Dokuchaev, V. V., 1883, *The Black Soil of Russia*. Published by the Imperial Economic Society, St Petersburg. (Russian)

Griendt, J. S. van de, 2009, *Lekkerkerk 30 jaar geleden*. Bouwfonds Ontwikkeling

Hartemink, A. E., ed., 2006, *The Future of Soil Science*. International Union of Soil Sciences, Wageningen

—, 2010, 'Edelman boor'. In: J. Bouma et al., eds, *Profiel van de Nederlandse Bodemkunde, 75 jaar Nederlandse Vereniging, 1935–2010*. Nederlandse Bodemkundige Vereniging, 45

Hauvette, H., 1921, *Dante. Inleiding tot de studie van de Divina Commedia*. Trans. William Davids. Wereldbibliotheek, Amsterdam

Lucian, *The Works*. Trans. H. W. Fowler and F. G. Fowler (n.d.). Accessed at ebooks.adelaide.edu.au

Moon, D., 2005, 'The Environmental History of the Russian Steppes: Vasilii Dokuchaev and the Harvest Failure of 1891'. *Transactions of the Royal Historical Society* 15, 149–174

Moormann, F. R., N. van Breemen, 1986, *Rice: Soil, Water, Land*. International Rice Research Institute, Los Baños, the Philippines

Pruissen, F.G.M. van, and B. W. Zuurdeeg, 1988, 'Hoge metaalgehalten in ijzeroerknollen in de Nederlandse bodem'. *Milieutechniek* 3, 84–91

Reden, S. von, 1997, 'Money, Law and Exchange: Coinage in the Greek Polis'. *The Journal of Hellenic Studies*, 117, 154–76

Riet, B. P. van de, E.C.H.E.T. Lucassen, R. Bobbink, J. H. Willems and J.G.M. Roelofs, 2005, *Preadvies Zinkflora*. Knowledge Division, Ministry of Agriculture, Nature and Food Quality, Ede

Stevens, S. T., 1991, 'Charon's Obol and Other Coins in Ancient Funerary Practice'. *Phoenix*, 45/3, 215–29

Stone, R., 2009, 'Archaeologists Seek New Clues to the Riddle of Emperor Qin's Terra-Cotta Army'. *Science*, 325, 22–3

Wolff, J., 2007, 'Emperor Qin in the Afterlife'. *Writing* 20, 10–16. Accessed at twp.duke.edu.

six: Charon's Ferry

Andel, Tj. J. Van, and C. N. Runnels, 2005, 'Karstic Wetland Dwellers of Middle Palaeolithic Epirus, Greece'. *Journal of Field Archaeology*, 30/4, 367–84

Aristophanes, *The Frogs*. Trans. Ian Johnston. Accessed at records.viu.ca

Bar-Matthews, M., A. Ayalon, M. Gilmour, A. Matthews and C. J. Hawkesworth, 2003, 'Sea-Land Oxygen Isotopic Relationships from Planktonic Foraminifera and Speleothems in the Eastern Mediterranean Region and their Implication for Paleorainfall during Interglacial Intervals'. *Geochimica et Cosmochimica Acta*, 67/17, 3181–99

Besonen, M. R., 1997, *The Middle and Late Holocene Geology and Landscape Evolution of the Lower Acheron River Valley, Epirus, Greece*. Thesis, University of Minnesota

—, G. Rapp and Z. Jing, 2003, 'The Lower Acheron River Valley: Ancient Accounts and the Changing Landscape'. *Hesperia Supplements, 32: Landscape Archaeology in Southern Epirus, Greece*, 1, 199–263

Bosence, D.W.J., and R.C.L. Wilson, 2003, 'Carbonate Depositional Systems'. In: A. Coe, ed., *The Sedimentary Record of Sea-Level Change*. The Open University/Cambridge University Press, 209–33

Clendenon, C., 2009, 'Karst Hydrology in Ancient Myths from Arcadia and Argolis, Greece'. *Acta Carsologica* 38/1, 145–54

Cvijić, J., 1893, *Das Karstphänomen*. Dissertation, Vienna

Ford, D., 2007, 'Jovan Cvijić and the Founding of Karst Geomorphology'. *Environmental Geology* 51, 675–684

Heggen, R.J., 2011. *Underground Rivers*. The University of New Mexico

Hesiod, *Theogony*. Trans. H. G. Evelyn-White (n.d.). Accessed at www.greekmythology.com

Hinsbergen, D.J.J. van, W. J. Zachariasse, M.J.R. Wortel and J. E. Meulenkamp, 2005, 'Underthrusting and Exhumation: A Comparison between the External Hellenides and the "Hot" Cycladic and "Cold" South Aegean Core Complexes (Greece)'. *Tectonics*, 24.

Kempe, S., and W. Rosendahl, 2008, *Höhlen, verborgene Welten*. Primus Verlag, Darmstadt

LaMoreaux, P. E., and J. W. LaMoreaux, 1998, 'A History of Karst Studies: From the Stone Age to the Present'. *Focus on Geography*, 45, 22–27

Leake, W. M., 1830, *Travels in the Morea* (three parts). John Murray, London

Marnelis, F., N. Roussos, N. Rigakis and V. Karakitsios, 2007, *Structural Geology of the Western Greece Fold and Thrust Belt. Guide to Fieldtrip No. 1*. AAPG and AAPG European Region Energy Conference and Exhibition 2007, Athens

Mousselimis, S., 1989, *The Ancient Underworld and the Oracle for Necromancy at Ephyra*. University of Ioannina

Pausanias, *Description of Greece*. Trans. W.H.S. Jones (n.d.). Accessed at www.theoi.com

Petrocheilou, A., 1984, *Les grottes de Grèce*. Ekdotike Athenon, SA, Athens

— (Petrochilou, A.), 1988, *La Grotte de Pérama a Ioannina*. Association Hellénique des Journalistes et Écrivains du Tourisme, Athens

Plato, *Phaedo*. Trans. Benjamin Jowett (n.d.). Accessed at ebooks.adelaide.edu.au

Ruddiman, W. F., 2001, *Earth's Climate: Past and Future*. W. H. Freeman & Co., New York

Runnels, C. N., and T. H. van Andel, 2003, 'The Early Stone Age of the Nomos of Preveza: Landscape and Settlement'. *Hesperia Supplements*, vol. 32, *Landscape Archaeology in Southern Epirus, Greece*, 1, 47–134

Seneca, n.d., *Hercules Furens*. Accessed at www.theoi.com

Sorel, D., 2000, 'A Pleistocene and Still Active Detachment Fault and the Origin of the Corinth-Patras Rift, Greece'. *Geology*, 28, 83–6

Strabo, 1927, *The Geography*, Book VIII. Loeb Classical Library Edition, Harvard University Press, Cambridge, MA

Thucydides, *History of the Peloponnesian War*. Trans. Richard Crawley. Accessed at people.ucalgary.ca

Turner, A. K., 1993, *The History of Hell*. Harcourt, Orlando, FL

Vassilopoulou, V., 2002, *The Cave of Lakes*. Ministry of Culture, Ephorate of Palaeoanthropology, Speleology, Athens

Westbroek, P., 1991, *Life as a Geological Force: Dynamics of the Earth*. W. W. Norton, New York

Wiseman, J., 1998, 'Rethinking the "Halls of Hades"'. *Archaeology*, 51/3, 12–18

SEVEN: Limbo

Alighieri, Dante, 1921, 'Convivio'. In: *Le Opere di Dante. Testo Critico della Societa Dantesca Italiana*, 145–315. Bemporad & Figlio, Florence

Aristotle, *On the Heavens*. Trans. J. L. Stocks (n.d.). Accessed at ebooks.adelaide.edu.au

Asín Palacios, M., 1919, *La Escatología Musulmana en la Divina Comedia*. Real Academia Española, imprenta de Estanislao Maestre, Madrid. Quote from *Islam and the Divine Comedy*. Trans. and abridged by Harold Sutherland. Accessed at books.google.nl

Bridoux, A., 1953, 'Introduction'. In: *Descartes, Oeuvres et Lettres*. Bibliothèque de la Pléiade, 9–23. Gallimard, Paris

Burnet, T., 1681, 1753, *The Sacred Theory of the Earth . . . in Four Books*. I: *Concerning the Deluge*. II: *Concerning Paradise*. III: *The Burning of the World*. IV: *The New Heavens and New Earth*. Glasgow

Cohen, F., 2007, *De herschepping van de wereld*. Uitgeverij Bert Bakker, Amsterdam

Descartes, R., 1647, 1989, *Oeuvres de Descartes, les Principes de la Philosophie*. Traduction française, IX 2. Facsimile uitgave. Librairie Philosophique J. Vrin, Paris

Galilée, 2008, *Leçons sur l'Enfer de Dante. (Due Lezioni all'Accademia Fiorentina circa la figura sito e grandezza dell'Inferno di Dante)*. Traduit par Lucette Degryse, postface de Jean-Marc Lévy-Leblond. Fayard, Paris

Gould, S. J., 1987, *Time's Arrow, Time's Cycle: Myth and Metaphor in the Discovery of Geological Time*. Harvard University Press, Cambridge, MA

Il libro della Scala di Maometto, 1991. Traduzione di Roberto Rossi Testa, a cura di Carlo Saccone. Conoscenza religiosa No. 7. Editore SE, Milan

Leonardo da Vinci, 1504–10, Codex Leicester. Accessed at www.odranoel.de

Pfister, L., H.H.G., Savenije and F. Fenicia, 2009, *Leonardo da Vinci's Water Theory*. International Association of Hydrological Sciences Special Publication 8, Wallingford, Oxfordshire

Robinson, T., 1694, *The Anatomy of the Earth*. Printed for J. Newton, London

—, 1696, *New Observations on the Natural History of the World of Matter*. London

Simoson, A. J., 2002, 'The Gravity of Hades'. *Mathematics Magazine*, 75, 5

—, 2007, *Hesiod's Anvil*. The Mathematical Association of America, Washington, DC

Steno, Nicola, 1669, *De solido intra solidum naturaliter contento dissertationis prodromus*. Florence

Tex, E. den, 1998, *Een voorspel van de moderne vulkaankunde in West-Europa met nadruk op de Republiek der Verenigde Nederlanden*. Koninklijke Nederlandse Akademie van Wetenschappen, Amsterdam

EIGHT: The City of Dis

Arsuaga, J. L., I. Martínez, A. Gracia, J. M. Carretero, C. Lorenzo and N. García, 1997, 'Sima de los Huesos (Sierra de Atapuerca, Spain): The Site'. *Journal of Human Evolution*, 33, 109–127

Cuenca-Bescós, G., J. Rofes and J. García-Pimienta, 2005, 'Environmental Change across the Early-Middle Pleistocene Transition: Small Mammalian Evidence from the Trinchera Dolina Cave, Atapuerca, Spain'. The Geological Society, London, Special Publications, 247, 277–86

Golany, G. S., 1992, *Chinese Earth-sheltered Dwellings: Indigenous Lessons for Modern Urban Design*. University of Hawaii Press

Hartmann, P., 2001, 'Die Entstehung des Valser Mineralwassers'. *Bulletin angewandter Geologie*, 6/1, 41–83

Le Pennec, J-L., A. Temel, J-L. Froger, S. Sen, A. Gourgaud and J-L. Bourdier, 2005, 'Stratigraphy and Age of the Cappadocia Ignimbrites, Turkey: Reconciling Field Constraints with Paleontologic, Radiochronologic, Geochemical and Paleomagnetic Data'. *Journal of Volcanology and Geothermal Research*, 141, 45–64

Luo, Wen-bao, 1987, 'Seismic Problems of Cave Dwellings on China's Loess Plateau'. *Tunnelling and Underground Space Technology*, 2/2, 203–8

Miskovsky, J-C., 1997, 'Paléoenvironnements de l'homme préhistorique d'après l'étude de karst et des remplissages de grottes et abris'. *Quaternaire*, 8/2–3, 319–27

Piedimonte, A. E., 2008, *Napoli sotterranea*. Edizioni Intra Moenia, Naples

Ren Meie, Liu Zhechun, Jin Jinluo, Deng Xiyang, Wang Feiyan, Peng Buzhou, Wang Xueyu and Wang Zonghan, 1981, 'Evolution of Limestone Caves in Relation to the Life of Early Man at Zhoukoudian, Beijing'. *Scientia Sinica*, 24/6, 843–51

Stevens, K., 1996, 'The Chinese Underworld and its Hierarchy'. In: R. Benewick and S. Donald, *Belief in China*. The Royal Pavilion, Art Gallery and Museums, Brighton

Tozzi, M., 2008, *Italia segreta*. Rizzoli, Milan

Voigt, S., A. Tetzlaff, Jianzhong Zhang, C. Künzer, B. Zhukov, G. Strunz, D. Oertel, A. Roth, P. van Dijk and H. Mehl, 2004, 'Integrating Satellite Remote Sensing Techniques for Detection and Analysis of Uncontrolled Coal Seam Fires in North China'. *International Journal of Coal Geology*, 59, 121–36

Yoon, Hong-key, 1990, 'Loess Cave-Dwellings in Shaanxi Province'. *GeoJournal*, 21/1–2, 95–102

Zhang, X., S. B. Kroonenberg and C. B. de Boer, 2004, 'Dating of Coal Fires in Xinjiang, Northwest China'. *Terra Nova*, 16, 68–74

—, J. L. van Genderen, H. Guan and S. B. Kroonenberg, 2003, 'Spatial Analysis of Thermal Anomalies from Airborne Multi-Spectral Data'. *International Journal of Remote Sensing*, 24/19, 3727–42.

NINE: Avarice

Agricola, G., 1950, *De re metallica*. Trans. H. C. Hoover and L. H. Hoover. Dover Publications, New York

Beaumont, P. B., 1973, 'The Ancient Pigment Mines of Southern Africa'. *South African Journal of Science*, 69, 140–46

Blunt, W., 1971, 2001, *Linnaeus, the Compleat Naturalist*. Frances Lincoln, London
Boomert, A., and S. B. Kroonenberg, 1977, 'Manufacture and Trade of Stone Artifacts in Prehistoric Surinam'. In: B.L. van Beek et al., eds: *Ex Horreo*. Albert Egges van Giffen Instituut voor Prae- en Protohistorie, Amsterdam, 9–44
Castro Caycedo, G. 2006, *El milagro de la Sal*. Bogotá
Crook, T. 1933, *History of the Theory of Ore Deposits*. Thomas Murby, London
Cunningham, C. G., R. E. Zartman, E. H. McKee, R. O. Rye, C. W. Naeser, O. Sanjinés, G. E. Ericksen and F. Tavera, 1996, 'The Age and Thermal History of Cerro Rico de Potosí, Bolivia'. *Mineralium Deposita*, 31, 374–85
D'Acosta, J., 1940, *Historia natural y moral de las Indias*. Fondo de cultura económica, México. English trans. accessed at books.google.nl
D'Arezzo, R., 1282, 1997, *La composizione del mondo*. Fondazione Pietro Bembo, Ugo Guanda Editore, Parma
Forbes, R. J., 1943, 'De antieke mijnbouw'. *Geologie en Mijnbouw*, 5, 1–2, 5–8
Het Gilgamesj-epos, 2008. Vertaling Theo de Feyter. Ambo, Amsterdam
Gowland, W., 1912, 'The Metals in Antiquity'. *The Journal of the Royal Anthropological Institute of Great Britain and Ireland*, 42, 235–87
Hoffmann, E.T.A., *The Mines of Falun*. No translator given. Accessed at www .horrormasters.com
Jesus, P. S. De, 1978. 'Metal Resources in Ancient Anatolia'. *Anatolian Studies*, 28, 97–102
Jones, J. E., 1982, 'The Laurion Silver Mines: A Review of Recent Researches and Results'. *Greece and Rome*, 2nd ser., 29/2, 169–83
Koark, H. J., 1962, 'Zur Altersstellung und Entstehung der Sulfiderze vom Typus Falun'. *Geologische Rundschau* 52, 124–46
Kroonenberg, S. B., 2001, 'Down to Earth'. *Jaarboek Mijnbouwkundige Vereniging*, 2000–2001, 55–9
Lagerlöf, S., n.d., *The Wonderful Adventures of Nils*. Trans. Velma Swanston Howard
Louwe Kooymans, L. P., P. W. van den Broeke, H. Fokkens and A. L. van Gijn, 2005, *Nederland in de prehistorie*. Uitgeverij Bert Bakker, Amsterdam
Lüschen, H., 1968, *Die Namen der Steine*. Ott Verlag, Thun
Moorey, P.R.S., 1994, *Ancient Mesopotamian Materials and Industries: The Archaeological Evidence*. Oxford University Press
Nriagu, J. O., 1983, 'Occupational Exposure to Lead in Ancient Times'. *The Science of the Total Environment*, 31, 105–16
Pastor Poppe, R., 1995, *Cuentos mineros del Siglo xx*. Editorial Los Amigos del Libro, Cochabamba, Bolivia
Polo, Marco, 2000, *Il milione*. Biblioteca Universale Rizzoli, Milan. Quotation from 'The Travels', Penguin Books, trans. R. E. Latham.
Rademakers, P.C.M., ed., 1998, *De prehistorische vuursteenmijnen van Ryckholt-St. Geertruid*. Werkgroep Prehistorische Vuursteenmijnbouw, Nederlandse Geologische Vereniging
Ramírez Velarde, F., 1988, *Socavones de angustia*. Los tiempos-Los Amigos del Libro, Cochabamba, Bolivia

Roever, W. P. de, Oen Ing Soen, 1970, *Programma van de geologische excursie naar Zuid-Zweden en Zuid-Noorwegen*. University of Amsterdam
Rothenberg, B., 1972, *Timna: Valley of the Biblical Copper Mines*. Thames and Hudson, London
Sözen, M., ed., 2000, *Cappadocia*. Ayhan Sahenk Foundation, Istanbul
Sundblad, K., 1994, 'A Genetic Reinterpretation of the Falun and Åmmeberg Ore Types, Bergslagen, Sweden'. *Mineralium Deposita*, 29, 170–79
Swedenborg, E., 1734, *Opera Philosophica Et Mineralia*. Friedrich Hekel, Leipzig. Quotation from 'Swedenborg's Treatise on Copper', *Nature*, 5 November 1938, accessed at books.google.nl
—, 1758, 2000, *De Caelo et Ejus Mirabilibus et de inferno. Ex Auditis et Visis*, Heaven and Hell. Trans. G. F. Dole. Swedenborg Foundation, West Chester, Pennsylvania
Tegengren, F. R., 1926, *Sveriges ädlare malmer och bergverk*. Sveriges Geologiska Undersökning, Stockholm
Theophrastus, 1956, *On Stones*. Introduction, Greek text, English translation and commentary by E. R. Caley and J.F.R. Richards. Ohio State University, Columbus
Ujueta, G. L., 1969, 'Salt in the Eastern Cordillera of Colombia'. *The Geological Society of America Bulletin*, 80, 2317–20
Werner, D., 1990, *Bergmannssagen aus dem sächsischen Erzgebirge*. Deutscher Verlag für Grundstoffindustrie GmbH, Leipzig
Willies, L., 1992, 'Report on the 1991 Archaeological Survey of Kestel Tin Mine,Turkey'. *Bulletin of the Peak District Mines Historical Society*, 11/5, 241–7
Yang, Hanchen, Yi Xianrui, Yi Shuangting, Song Jianzhong and Min Yaoming, 1986, *Xinjiang Gems and Jades*. Xinjiang People's Publishing House, Urumqi
Yener, K. A., H. Özbal, E. Kaptan, A. N. Pehlivan and M. Goodway, 1989, 'Kestel: An Early Bronze Age Source of Tin Ore in the Taurus Mountains, Turkey'. *Science*, 244, 200–03

TEN: The Conflagration

Haakman, A., 1991, *De onderaardse wereld van Athanasius Kircher*. Meulenhoff, Amsterdam
Kircher, A., 1669, *The Volcanoes, or Burning and Fire Vomiting Mountains, Famous in the World*. Translator not named. Kessinger, Whitefish, Montana
Richter, D., 2007, *Der Vesuv: Geschichte eines Berges*. Verlag Klaus Wagenbach, Berlin
Walker, D. P., 1964, *The Decline of Hell*. Routledge and Kegan Paul, London

ELEVEN: The Monster Geryon

Baker, B. J., et al., 2003, 'Related Assemblages of Sulphate-Reducing Bacteria Associated with Ultradeep Gold Mines of South Africa and Deep Basalt Aquifers of Washington State'. *Environmental Microbiology*, 5/4, 267–77

Burroughs, E., 1922, *At the Earth's Core*. Wildside Press, Doylestown, PA
Casanova, G., 1788, 1967, *Icosaméron*. Éditions François Bourin, Paris
Galilei, G., 1632, 1877 *Dialogo sopra i due massimi sistemi*. Edoardo Sonzogno Editore, Milan
—, 1638, 1990, *Discorsi e Dimostrazioni matematiche intorno a due Nuove Scienze*. Einaudi, Turin
Gjevik, B., H. Moe and A. Ommundsen, 1997, 'Sources of the Maelstrom'. *Nature* 388/28, 837–8
Gordon, I., 1958, 'A New Subterranean Crustacean from the West Indies'. *Nature*, 181, 1552–3
Griffin, D. A., 2004, 'Hollow and Habitable Within: Symmes's Theory of Earth's Internal Structure and Polar Geography'. *Physical Geography*, 25/5, 382–97
Halley, E., 1692, 'An Account of the Cause of the Change of the Variation of the Magnetical Needle; With an Hypothesis of the Structure of the Internal Parts of the Earth'. *Philosophical Transactions*, XVI (1692), 563–87
Holthuis, L. B., 1978, 'Zoological Results of the British Speleological Expedition to Papua New Guinea 1975: 7 Cavernicolous Shrimps (Crustacea, Decapoda, Natantia) from New Ireland and the Philippines'. *Zoölogische Mededelingen*, 53/19, 209–24
Kollerstrom, N., 1992, 'The Hollow World of Edmond Halley'. *Journal for the History of Astronomy*, 23, 185–92
Kranjc, A., 2006, 'Seasonal Karst Lake Cerknica (Slovenia): 2000 Years of Man Versus Nature'. *Helictite*, 39/2, 39–46
Manning, R. B., C. W. Hart and T. M. Iliffe, 1986, 'Mesozoic Relicts in Marine Caves of Bermuda'. *Stygologia*, 2/1–2, 156–66
—, and L. B. Holthuis, 1984, '*Geryon fenneri*, a New Deepwater Crab from Florida (Crustacea: Decapoda: Geryonidae)'. *Proceedings of the Biological Society of Washington*, 97/3, 666–73
Newton, I., 1687, *Philosophiae naturalis principia mathematica*, London
Noordijk, J., R.M.J.C. Kleukers, F. J. van Nieukerken and A. J. van Loon, ed., 2010, *De Nederlandse biodiversiteit*. Nederlands Centrum voor Biodiversiteit Naturalis, Leiden
Pedersen, K. 1993, 'The Deep Subterranean Biosphere'. *Earth-Science Reviews*, 34, 243–60
Poe, E. A., 1841, 1941, 'A Descent into the Maelström'. In: *Tales of Mystery and Imagination*. The Heritage Press, New York
Sket, B., 1999, 'The Nature of Biodiversity in Hypogean Waters and How It Is Endangered'. *Biodiversity and Conservation*, 8, 1319–38
Symmes, A., 1885, *The Symmes Theory of Concentric Spheres: Demonstrating that the Earth is Hollow, Habitable Within, and Widely Open About the Poles*. Bradley and Gilbert, Louisville, Kentucky
Szewzyk, U., R. Szewzyk and T.-A. Stenström, 1994, 'Thermophilic, Anaerobic Bacteria Isolated from a Deep Borehole in Granite in Sweden'. *Proceedings of the National Academy of Sciences*, 91, 1810–13
Voltaire, 1752, 2000, *Micromégas*. Le Livre de Poche, Paris

Westbroek, P., 2009, *Terre! Des menaces globales à l'espoir planétaire.* Éditions du Seuil, Paris

Yaldwyn, J. C., 1960, 'Crustacea Decapoda Natantia from the Chatham Rise: A Deep Water Bottom Fauna from New Zealand. Biological Results of the Chatham Island 1954 Expedition'. *New Zealand Department of Scientific and Industrial Research Bulletin*, 139, 13–53

Zoback, M. D., and R. Emmermann, 1994, *Scientific Rationale for Establishment of an International Program of Continental Scientific Drilling.* International Lithosphere Program, Potsdam

TWELVE: The River of Tar

Buhler, R., 1998, 'Background on the Drilling to Hell Story'. Accessed at www.truthorfiction.com

Dumas, A., 2002, *Voyage au Caucase.* Hermann, Éditeurs des sciences et des arts, Paris

Elkind, P., and D. Whitford, 2011, 'An Accident Waiting to Happen'. *Fortune*, 7 February, 2011, 53–71

Green, T., N. Abdullayev, J. Hossack, G. Riley and A. M. Roberts, 2009, 'Sedimentation and Subsidence in the South Caspian Basin, Azerbaijan'. In: M.-F. Brunet, M. Wilmsen and J. W. Granath, eds, *South Caspian to Central Iran Basins.* The Geological Society, Special Publications, 312, 241–60

Guliyev, I., A. A. Feyzullayev, 1997, *All About Mud Volcanoes.* Geology Institute, Azerbaijan National Academy of Sciences, Baku, Azerbaijan

Hanway, J., 1753, *A Historical Account of the British Trade over the Caspian Sea.* London

Head, I. M., D. M. Jones and S. R. Larter, 2003, 'Biological Activity in the Deep Subsurface and the Origin of Heavy Oil'. *Nature Publishing Group*, 426, 344–52

Heward, A. P., S. Chuenbunchom, G. Mäkel, D. Marsland and L. Spring, 2000, 'Nang-Nuan oil field, 6/27, Gulf of Thailand: Karst Reservoirs of Meteoric or Deep-Burial Origin?' *Petroleum Geoscience*, 6, 15–27

Institute of the History of Natural Sciences, Chinese Academy of Sciences, 2009, *Ancient China's Technology and Science.* Foreign Language Press, Beijing

Kazansky, V. I., K. V. Lobanov and N. V. Sharov, 2007, 'From the Kola Superdeep Borehole Section towards 3D Models of the Pechenga Ore District: 10th Anniversary of the Discovery no 28 in Earth Sciences'. *Bulletin of the Russian Academy of Natural Sciences*, 2, 3–7

Knott, T., 2006. 'When the Pendulum Swings'. *Frontiers* (BP), December 2006, 19–25

Kroonenberg, S. B., 2002. 'Vulkanisch modderballet'. *Natuur en Techniek*, 70/9, 54–62

—, M. D. Simmons, N. I. Alekseevski, E. Aliyeva, M. B. Allen, D. N. Aybulatov, A. Baba-Zadeh, E. N. Badyukova, C. E. Davies, D. J. Hinds, R. M. Hoogendoorn, D. Huseynov, B. Ibrahimov , P. Mamedov, I. Overeem, G. V. Rusakov†, S. Suleymanova, A. A. Svitoch and S. J. Vincent, 2005, 'Two Deltas, Two Basins, One River, One Sea: The Modern Volga Delta as an Analogue of the Neogene

Productive Series, South Caspian Basin'. In: L. Giosan and J. Bhattacharya, eds, *River Deltas: Concepts, models and examples*. Special Volume SEPM 83, 231–56
Leeuw, C. van der, 2000, *Oil and Gas in the Caucasus and Caspian: A History*. Curzon, Richmond, UK
Mirzoev, M. A., 1995, *On the Shelf of the Caspian Sea*. Sabach, Baku, Azerbaijan (Russian)
Vetrin, V. R., O. M. Turkina and Ø. Nordgulen, 2001, 'Surface Analogues of "Grey Gneiss" among the Archaean Rocks in the Kola Superdeep Borehole. (Experiences from Petrological-Geochemical Modelling of Lower Crust Composition and Conditions of Formation of Tonalite-Trondhjemite Rocks)'. *Russian Journal of Earth Sciences*, 3/3, 136–69
Vogel, H. U., 1993, 'The Great Well of China'. *Scientific American*, June 1993, 116–21
Yakubov, A. A., and A. A. Alizadeh, 1971, *Atlas of Mud Volcanoes in Azerbaijan*. Azerbaijan Academy of Sciences, Baku (in Russian).

THIRTEEN: Collapses

Berendsen, H.J.A., B. Makaske, O. van de Plassche, M.H.M. van Ree, S. Das, M. van Dongen, S. Ploumen and W. Schoenmakers, 2007, 'New Groundwater-Level Rise Data from the Rhine-Meuse Delta: Implications for the Reconstruction of Holocene Relative Mean Sea-Level Rise and Differential Land-Level Movements'. *Netherlands Journal of Geosciences – Geologie en Mijnbouw*, 86/4, 333–54
Bouvard, J., 2005, *Le métro de Moscou*. Éditions du Sextant, Paris
Diffre, P., and C. Pomerol, 1979, *Paris et Environs, Les roches, l'eau et les hommes. Guides géologiques régionaux*. Masson, Paris
Gaffard, E., 2007, *Paris souterrain*. Parigramme, Paris
Gans, W. de, 2006, *Geologieboek Nederland*. ANWB, Netherlands
Jelgersma, S., 1961, *Holocene Sea-Level Changes in the Netherlands*. Mededelingen van de Geologische Stichting, Serie, c6 (7). Haarlem
Kroonenberg, S. B., 2010. *Onder de groene zoden begint het pas*. Farewell speech, Delft Technical University
Lomonosov, M. V., 1763, 1949, *About the Layers of the Earth*. Gosgeoizdat, Moscow. (Russian)
Plassche, O. van de, 1982, *Sea-Level Change and Water-Level Movements in the Netherlands during the Holocene*. Mededelingen Rijks Geologische Dienst 36, Haarlem

FOURTEEN: The Lead Cloak

Bijwaard, H., and W. Spakman, 1999, 'Tomographic Evidence for a Narrow Whole Mantle Plume Below Iceland'. *Earth and Planetary Science Letters*, 166, 121–6
Brey, G. P., V. Bulatov, A. Girnis, J. W. Harris and T. Stachel, 2004, 'Ferropericlase: A Lower Mantle Phase in the Upper Mantle'. *Lithos*, 77, 655–63
Chung, H. Y., and J. E. Mungall, 2009, 'Physical Constraints on the Migration of

Immiscible Fluids through Partially Molten Silicates, with Special Reference to Magmatic Sulfide Ores'. *Earth and Planetary Science Letters*, 286, 14–22

Conrad, V., 1925, *Laufzeitkurven des Tauernbebens vom 28.November 1923*. Mitteilungen der Erdbeben-Kommission, Neue Folge Nr. 59, Österr. Akad. d. Wissenschaften, Vienna

Courtillot, V., 2009, *Nouveau voyage au centre de la terre*. Odile, Jacob, Paris

—, A. Davaille, J. Besse and J. Stock, 2003, 'Three Distinct Types of Hotspots in the Earth's Mantle'. *Earth and Planetary Science Letters*, 205, 295–308

Dasgupta, R., and M. M. Hirschmann, 2010, 'The Deep Carbon Cycle and Melting in Earth's Interior'. *Earth and Planetary Science Letters*, 298, 1–13

Guðmundsson, A. T., 2007, *Living Earth: Outline of the Geology of Iceland*. Mál og Menning, Reykjavik

Holmes, A., 1965, *Principles of Physical Geology*. Nelson, London

Holub, E., 1881, *Sieben Jahre in Süd-Afrika*. Alfred Holder, Vienna.

Kroonenberg, S. B., 1976, *Amphibolite Facies and Granulite-Facies Metamorphism in the Coeroeni-Lucie Area, Southwestern Surinam*. PhD thesis, University of Amsterdam

Meijer, R.J. de, and W. van Westrenen, 2009, *Hoe werkt de aarde?* Wetenschappelijke Bibliotheek van Natuurwetenschap en Techniek, Diemen

Mohorovičić, A., 1910, 1992. 'Earthquake of 8 October 1909'. Trans. in *Geofizika*, 9–55

Munk, W., 1982, 'Affairs of the Sea'. In: *A Celebration of Geophysics and Oceanography*. Scripps Institution of Oceanography Reference Series, 84–5, La Jolla, California

Nemchin, A. A., M. J. Whitehouse, M. Menneken, T. Geisler, R. T. Pidgeon and S. A. Wilde, 2008, 'A Light Carbon Reservoir Recorded in Zircon-Hosted Diamond from the Jack Hills'. *Nature*, 454, 92–5

Press, F., and R. Siever, 2004, *Understanding Earth*, 2nd edn. Freeman, New York

Roberts, B., 1976, 1984, *Kimberley, Turbulent City*. David Philip Publishers, Claremont, South Africa

Roever, W. P. de, 1957, 'Sind die alpinotypen Peridotit-massen vielleicht tektonisch verfrachtete Bruchstücke der Peridotitschale?' *Geologische Rundschau*, 46/1, 137–46

Zheng, Y. F., 1992, 'Sulfur Isotopes in the Mantle'. *Naturwissenschaften*, 79, 511–12

FIFTEEN: To the Centre of the Earth

Chojecki, C. E., 1857, *Voyage dans les mers du Nord à bord de la corvette la Reine Hortense*. Michel Lévy Frères, Paris

Dupuy, L., 1999, *Espace et temps dans l'oeuvre de Jules Verne. Voyage au centre de la terre . . . et dans le temps*. Dissertation for Certificat International d'Écologie Humaine. Université de Pau et des Pays de l'Adour, France

—, 2009, *Géographie et imaginaire géographique dans les Voyages Extraordinaires de Jules Verne: Le Superbe Orénoque (1898)*. PhD thesis, Université de Pau et des Pays de l'Adour, France

Gaimard, P., 1840, *Voyage en Islande et au Groënland executé pendant les années 1835 et 1836 sur la corvette la Recherche, commandée par m. Tréhouart, dans le but de découvrir les traces de la Lilloise.* Arthus Bertrand, Paris
Gutenberg, B., 1964, 'Low-Velocity Layers in the Earth's Mantle.' *Bulletin of the Geological Society of America*, 65, 337–348
Robert, E., 1840, *Géologie et minéralogie. Atlas.* In: P. Gaimard, *Voyage en Islande et au Groënland.* Arthus Bertrand, Paris
Sand, G., 1864, 2011, *Laura, Voyage dans le cristal.* Dodo Press, Gloucester
Umbgrove, J.H.F., 1942, *The Pulse of the Earth.* Martinus Nijhoff, The Hague
Verne, J., 2006, *The Extraordinary Journeys: Journey to the Centre of the Earth.* Trans. William Butcher. Oxford University Press

SIXTEEN: With Lucifer in the Ice

Lehmann, I., 1936, 'P''. *Publication Bureau Central Séismologique Internationale, Travaux Scientifiques*, Série A, 14, 87–115
—, 1987, 'Seismology in the Days of Old'. *Eos*, 68/3, 33–5
Oldham, R. D., 1906, 'The Constitution of the Interior of the Earth, as Revealed by Earthquakes'. *Quarterly Journal of the Geological Society*, 62, 456–75
Plutarch, 1878, *The Vision of Aridaeus*. Trans. W. W. Goodwin in: *Moralia*, 566f; Little, Brown and Company, Boston
Stevenson, D. J., 2003. 'Mission to Earth's Core: A Modest Proposal'. *Nature*, 423, 239–40
Wiechert, E., 1897, 'Über die Massenverteilung im Inneren der Erde'. *Nachrichten der Königlichen Gesellschaft der Wissenschaften zu Göttingen, Mathematisch-Physikalische Klasse*, 221–43

SEVENTEEN: The Way Back

Fyfe, W. S., V. Babuska, N. J. Price, E. Schmid, C. F. Tsang, S. Uyeda and B. Velde, 1984, 'The Geology of Nuclear Waste Disposal'. *Nature*, 310, 537–40
Song, X., 1997, 'Anisotropy of the Earth's Inner Core'. *Reviews of Geophysics*, 35/3, 297–313
—, and P. G. Richards, 1996, 'Seismological Evidence for Differential Rotation of the Earth's Inner Core'. *Nature*, 382, 221–4
Vian, B., 1947, *L'Écume des jours*. Gallimard, Paris

Acknowledgements

This book would not have become what it is without the support and advice of those who read the manuscript and offered their comments. Frank Westerman, my mentor from the very beginning, helped me to be stricter in my choices. With her implacable logic, my tireless editor Jessica Nash sometimes asked me such astute questions that I had no answer for them. Anita Roeland not only sifted out a shamefully large quantity of careless errors, but also contributed her broad knowledge of the classics. I could always rely on Mariette Blokhuis to come up with radical solutions. Anton Haakman was on my wavelength and referred to interesting supplementary literature. The book was also improved by the subtle literary instinct of my son Sal Kroonenberg. Thanks to you all. I also thank my Dutch publishers Emile Brugman and Erna Staal for their support and enthusiasm. I am immensely grateful to Andy Brown for his meticulous and resourceful translation into English. It was a great pleasure to work so closely together with him.

In addition, I am very grateful to all those who helped to obtain information, including Stefan Bergman, Joy Burrough, Raymond J. Clark, Floris Cohen, Dick Delforterie, Hans Diederix, Theo Gerritsen, Alet Heezemans, Hester Helming, Michail Lychagin, Silvana Russo, Huub Savenije, Carolina Sigarán, Erki Tammiksaar, Maaike van Tooren, Rosario Varriale and Ger Wieberdink. The line drawings were made by Ton Markus of Kartomedia. He was very patient with me.

Photo Acknowledgements

The Author and publishers wish to express their thanks to the below sources of illustrative material and / or permission to reproduce it.

Photos courtesy of the author: pp. 8, 16, 19, 29, 35, 38, 46, 53, 59, 68, 84, 87, 90, 92, 97, 99 (top), 101 (bottom), 108 54.504 mmtop), 113, 136, 140, 142, 143, 144, 146, 153, 163, 164 (left and right), 165, 167, 168, 175, 176, 179, 204 (top), 209 (top and bottom), 212 (top and bottom), 213 (bottom), 215, 227, 231, 232, 235, 238, 240 (bottom), 243, 256, 257, 258 (top), 262, 268, 279.

After www.4dimensions.org: p. 78 (right); from G. Agricola, *De re metallica* [1556] (New York, 1950): p. 160; from A. Avallone, P. Briole, C. Delacourt, A. Zollo and F. Beauducel, 'Subsidence of Campi Flegrei (Italy) detected by SAR interferometry', *Geophysical Research Letters*, XXVI/15 (1999): p. 48; © BP plc: p. 221; © Iwan Baan: p. 147; after M. Bar-Matthews et al., 'Sea–Land Oxygen Isotopic Relationships from Planktonic Foraminifera and Speleothems in the Eastern Mediterranean Region and their Implication for Paleorainfall During Interglacial Intervals', *Geochimica et Cosmochimica Acta*, LXVII/17 (2003): p. 99 (bottom); after M. R. Besonen, *The Middle and Late Holocene Geology and Landscape Evolution of the Lower Acheron Valley, Epirus, Greece* (thesis, University of Minnesota, 1997): p. 88; Biblioteca Apostolica Vaticana, Rome: p. 112; © Bijian: p. 133; © Mariette Blokhuis: p. 148; after D.W.J. Bosence and R.C.L. Wilson, 'Carbonate Depositional Systems', in *The Sedimentary Record of Sea-level Change*, ed. A. L. Coe (Milton Keynes, 2003): p. 106; from Thomas Burnet, *The Sacred Theory of the Earth . . .* (Glasgow, 1753): p. 129; photo © Douglas Campbell: p. 173; © ciaas.no: p. 225; © Conocphillips.no: p. 225; photo Corbis: p. 125; after V. Courtillot et al., 'Three Distinct Types of Hotspot in the Earth's Mantle', *Earth and Planetary Science Letters*, 205 (2003): p. 247; after René Descartes, *Les Principes de la philosophie* [1647] (Paris, 1989): p. 127; DinoLoket/TNO: p. 11; École des Mines, Paris: p. 103 (top and bottom); photo courtesy Earth Science Museum, Moscow State University: p. 63; from B. G. Escher, *De methodes der grafische voorstelling . . .* (Amsterdam, 1934): p. 271; Bill Gates collection: p. 125; Google Earth: p. 26; from W. Gowland, 'The Metals in Antiquity', *The Journal of the Royal Anthropological Institute of Great Britain and Ireland*, 42 (1912): p. 157 (top); from D. A. Griffin, 'Hollow and Habitable Within: Symmes's Theory of Earth's Internal Structure and Polar Geography', *Physical Geography*, XXV/5 (2004): p. 197; after A. T. Guðmundsson, *Living Earth:*

Outline of the Geology of Iceland (Reykjavik, 2007): p. 244; from William Hamilton, *Campi Flegrei: Osservazioni sui vulcani delle Due Sicilie* (Naples, 1776): p. 32; after Henri Hauvette, *Dante: Inleiding tot de Studie van de Divina Commedia* (Amsterdam, 1921): pp. 73, 229; after D.J.J. van Hinsbergen, W. J. Zachariasse, M.J.R. Wortel and J. E. Meulenkamp, 'Underthrusting and Exhumation: A Comparison between the External Hellenides and the 'Hot' Cycladic and 'Cold' South Aegean Core Complexes (Greece)', *Tectonics*, XXIV/2 (2005): p. 85; after A. Holmes, *Principles of Physical Geography* (London, 1965): p. 239; from E. Holub, *Sieben Jahre in Süd-Afrika* (Wenen, 1881): p. 249 (top); J. G. Howes: p. 15; © Bjorn Hroarsson/ Extreme Iceland: p. 259; from A. A. Jakoebev and A. A. Alizadeh, *Atlas van Moddervulkanen in Azerbeidjan* (Bakoe, 1971): p. 211; from Andrea de Jorio, *Viaggio di Enea all'inferno, ed agli elisii secondo Virgilio* (Naples, 1825): p. 80; © Kartomedia: pp. 9, 15, 50, 75, 77, 78 (right), 79, 85, 88, 93, 98, 99, 104, 122, 183, 218, 239, 272; after S. Kempe and W. Rosendahl, *Höhlen: Verborgene Welten* (Darmstadt, 2008): p. 98; from Athanasius Kircher, *d'Onder-aardse weereld* [translation of *Mundus subterraneus, in XII libros digestus . . .*] (Amsterdam, 1664): pp. 43, 190, 192; after S. B. Kroonenberg, *Amphibolite Facies and Granulite-facies Metamorphism in the Coeroeni-Lucie Area, Southwestern Surinam* (thesis, University of Amsterdam, 1976): p. 241; from Clarence Larkin, *Rightly Dividing the World* (Glenside, PA, 1920): p. 20; © Carlos Lazcano Sahagún Cuando: p. 261; from L. P. Louwe Kooymans et al., *Nederland in de Prehistorie* (Amsterdam, 2005): p. 150; photo © Erich Lessing/ Hollandse Hoogte: p. 196; photos © M. Lytsjagin: pp. 60, 81, 224; from R. B. Manning and L. B. Holthuis, '*Geryon fenneri*, a New Deepwater Crab from Florida (Crustacea: Decapoda: Geryonidae)', *Proceedings of the Biological Society of Washington*, XCVII/3 (1984): p. 200; 'Master of Delft' (1480–1500) © Rijksmuseum, Amsterdam: p. 13; after R. J. de Meijer and W. van Westrenen, *Hoe Werkt de Aarde* (Diemen, 2009): p. 246; Mineralological and Geological Museum, Technische Universiteit Delft: p. 262; Museo del Prado, Madrid: p. 93 (bottom); National Museum of Iceland, Reykjavik: p. 108 (top); Natural History Museum, London: p. 105; © Noord/Zuidlijn: p. 230; photo courtesy Odessa Numismatics Museum: p. 69; after R. D. Oldham, 'The Constitution of the Interior of the Earth, as Revealed by Earthquakes', *Quarterly Journal of the Geological Society*, LXII/106: p. 270; after L. Pfister, H.H.G. Savenije and F. Fenicia, *Leonardo da Vinci's Water Theory* (Wallingford, 2009): p. 126; © Neil Piggott: p. 213 (top); after F. Press and R. Siever, *Understanding Earth* (New York, 2004): pp. 242, 272; from E. Robert, 'Géologie et Minéralogie: Atlas', in *Voyage en Islande et au Groënland*, ed. P. Gaimard (Paris, 1840): pp. 254, 258 (bottom); from B. Roberts, *Kimberley: Turbulent City* (Kimberley, 1984): p. 248; after T. Robinson, *The Anatomy of the Earth* (London, 1694): p. 130; © W. P. de Roever: p. 240 (top); after W. F. Ruddiman, *Earth's Climate, Past and Future* (New York, 2001): p. 104; photo courtesy Sanderus Antiquariaat, Ghent: p. 42; C. W. Scott-Giles, illustration for Dorothy L. Sayers's translation of Dante's *Inferno*: pp. 108 (bottom), 114; from William Smith and George Grove, eds, *An Atlas of Ancient Geography, Biblical and Classical . . .* (London, 1874): p. 25; from M. Sözen, ed., *Cappadocia* (Istanbul, 2000): pp. 152, 155; http://spazioinwind.libero.it/pdf/mys1.pdf: p. 202; from Niels Stensen [Nicolaus Steno], *De solido intra solidum naturaliter contento dissertationis prodromus* (Florence, 1669): p. 128; photo Alex Suvorov/Fotolia: p. 64; Technische Universiteit Delft: p. 220 (bottom); after E. den Tex, *En Voorspel van de Moderne Vulkaankunde in West-Europa met Nadruk op de Republiek der Verenigde Nederlanden* (Amsterdam, 1998): p. 122; after http://thanasis.com: p. 79; after Ulrich Tichy (Wikimedia Commons): p. 101 (top); after Alice K. Turner, *The History of Hell*: p. 78 (left); after V. Vassilopolou, *The Cave of Lakes* (Athens, 2002): p. 93 (top); from J. Verne, *Voyage au centre de la terre* [1864] (Paris, 1912): pp. 255, 260; from M. di Vito, L. Lirer, G. Mastrolorenzo and G. Rolandi, 'The 1538 Monte Nuovo Eruption (Campi Flegrei, Italy)', *Bulletin of Vulcanology*, XLIX/15 (1987): p. 56; photo © Vox 2010–11: p. 100; Weekly World News (24 April 1990): p. 225 (bottom); after X. Zhang, J. L. van Genderen, H. Guan and S. B. Kroonenberg, 'Spatial Analysis of Thermal Anomalies from Airborne Multi-spectral Data', *International Journal of Remote Sensing*, XIV/19 (2003): p. 141; from M. D. Zoback and R. Emmermann, *Scientific Rationale for Establishment of an International Program of Continental Scientific Drilling* (Potsdam, 1994): p. 204 (bottom).

Index